MODULAR SERIES ON SOLID STATE DEVICES
Robert F. Pierret and Gerold W. Neudeck, Editors

Volume VI

Advanced Semiconductor Fundamentals

Second Edition

Robert F. Pierret
Purdue University

 Pearson

Harlow, England • London • New York • Boston • San Francisco • Toronto • Sydney
Dubai • Singapore • Hong Kong • Tokyo • Seoul • Taipei • New Delhi
Cape Town • São Paulo • Mexico City • Madrid • Amsterdam • Munich • Paris • Milan

Library of Congress Cataloging-in-Publication Data on file

Vice President and Editorial Director, ECS: *Marcia J. Horton*
Publisher: *Tom Robbins*
Editorial Assistant: *Jodi McDonnell*
Vice President and Director of Production and Manufacturing, ESM: *David W. Riccardi*
Executive Managing Editor: *Vince O'Brien*
Managing Editor: *David A. George*
Production Editor: *Scott Disanno*
Director of Creative Services: *Paul Belfanti*
Creative Director: *Carole Anson*
Art Director: *Jayne Conte*
Cover Designer: *Bruce Kenselaar*
Art Editor: *Greg Dulles*
Manufacturing Manager: *Trudy Pisciotti*
Manufacturing Buyer: *Lisa McDowell*
Marketing Manager: *Holly Stark*

© 2003 by Pearson Education, Inc.
Pearson Education, Inc.
Upper Saddle River, NJ 07458

Printed in the United States of America
47 2023

ISBN 0-13-061792-X

Pearson Education Ltd., *London*
Pearson Education Australia Pty. Ltd., *Sydney*
Pearson Education Singapore, Pte. Ltd.
Pearson Education North Asia Ltd., *Hong Kong*
Pearson Education Canada, Inc., *Toronto*
Pearson Educacìon de Mexico, S.A. de C.V.
Pearson Education—Japan, *Tokyo*
Pearson Education Malaysia, Pte. Ltd.
Pearson Education, Inc., *Upper Saddle River, New Jersey*

To my wife, Linda Pierret, and
to the memory of my mother, Elsie Pierret

Preface

Advanced Semiconductor Fundamentals is viewed by the author as a doorway to the graduate or journal-level discussion of solid-state devices. It was originally prepared in part as a supplement to a widely used graduate text and in part to provide background information required in advanced-level volumes of the Modular Series on Solid State Devices. Since its introduction in 1987, the volume has subsequently become routinely employed in introductory graduate-level courses on solid-state devices. The second edition primarily revises dated sections of the volume and, with a significant increase in end-of-chapter problems, expands its usefulness as a stand-alone text.

The designation "advanced" used in the title of the volume is of course a relative term: the material in the volume is "advanced" relative to that in Modular Series Volume I and chapters one through three in *Semiconductor Device Fundamentals*, other works by the author. The cited works are recommended prerequisites for the present volume. The present volume extends and reinforces the concepts presented in the cited works.

Following the general philosophy of the Modular Series, the present volume is devoted to a specific topic area and is essentially self-contained. The modular nature of the series permits the volumes to be used in courses of either standard or nonstandard format, the latter including short courses, television or web-based courses, and in-house continuing education courses. Students, practicing engineers, and scientists should also find this and the other volumes useful for individual instruction, whether it be for learning, reference, or review. Coherent presentation of the material in *Advanced Semiconductor Fundamentals* in the standard lecture format requires at least 15 fifty-minute periods. With minor deletions, the material in this volume is regularly covered during the first six weeks of a one-semester, three-credit-hour, first-year graduate-level course in Electrical and Computer Engineering at Purdue University.

The topic coverage in the second edition is essentially identical to that in the first edition. The treatment includes basic semiconductor properties, elements of Quantum Mechanics, energy band theory, equilibrium carrier statistics, recombination–generation processes, and drift/diffusion carrier transport. Unfortunately, length limitations precluded coverage of a number of other desirable topics. Nevertheless, the coverage should be sufficient for understanding or delving deeper into the operation of the major semiconductor device structures. Of the many semiconductors, silicon (Si) totally dominates the present marketplace; the vast majority of discrete devices and integrated circuits are silicon based. Given its position of dominance, attention is focused herein on Si in the text development. Where feasible, however, GaAs and other semiconductors are featured as the discussion warrants.

It should be mentioned that throughout the volume every effort has been made to use normally encountered symbols for a given quantity. In some instances this has led to dual-meaning symbols (e.g., k for wavenumber and for the Boltzmann constant). The proper interpretation of a dual-meaning symbol is invariably obvious from context. In the author's opinion it is preferable to court ambiguity rather than introduce alternative symbols and/or cumbersome subscripts that are unlikely to be encountered in other works.

Finally, I would like to acknowledge the influence of the classic text by McKelvey (J. P. McKelvey, *Solid State and Semiconductor Physics*, Harper and Row, New York, 1966). Chapter 2 and portions of Chapter 3 parallel McKelvey's organization and/or topic presentation. I would also like to gratefully acknowledge the assistance of Prof. Mark Lundstrom, a Purdue University colleague, who was most helpful in supplying key information on several topics and Tom Robbins, ECE Publisher at Prentice Hall, who exhibited great patience in dealing with a difficult author.

Prof. Robert F. Pierret
School of Electrical and Computer Engineering
Purdue University
W. Lafayette, IN

Contents

CHAPTER 1

Basic Semiconductor Properties

This chapter provides a brief introduction to semiconductors and semiconductor physics by surveying a select number of basic physical properties. It is the first step in building up the knowledge and analytical base required in the operational modeling of semiconductor devices. Major emphasis is placed on the structural description of materials in general and semiconductors in particular. Crystal structure is of central interest because it is intimately tied to the intrinsic electrical properties exhibited by a material.

1.1 GENERAL MATERIAL PROPERTIES

Table 1.1 lists the atomic compositions of semiconductors that are likely to be encountered in the device literature. As noted, the semiconductor family of materials includes the elemental semiconductors Si and Ge, compound semiconductors such as GaAs and ZnSe, and alloys like $Al_xGa_{1-x}As$.[†] Due in large part to the advanced state of its fabrication technology, Si is far and away the most important of the semiconductors, totally dominating the present commercial market. The vast majority of discrete devices and integrated circuits (ICs) including the central processing unit (CPU) in microcomputers and the ignition module in modern automobiles, are made from this material. GaAs, exhibiting superior electron transport properties and special optical properties is employed in a significant number of applications ranging from laser diodes to high-speed ICs. The remaining semiconductors are utilized in "niche" applications that are invariably of a high-speed, high-temperature, or optoelectronic nature. Given its present position of dominance, we will tend to focus our attention on Si in the text development. Where feasible, however, GaAs and other semiconductors will be featured as the discussion warrants.

[†] The x (or y) in alloy formulas is a fraction lying between 0 and 1. $Al_{0.3}Ga_{0.7}As$ would indicate a material with 3 Al atoms and 7 Ga atoms per every 10 As atoms.

Table 1.1   Semiconductor Materials

General Classification	Symbol	Semiconductor Name
(1) Elemental	Si	Silicon
	Ge	Germanium
(2) Compounds		
(a) IV–IV	SiC	Silicon carbide
(b) III–V	AlP	Aluminum phosphide
	AlAs	Aluminum arsenide
	AlSb	Aluminum antimonide
	GaN	Gallium nitride
	GaP	Gallium phosphide
	GaAs	Gallium arsenide
	GaSb	Gallium antimonide
	InP	Indium phosphide
	InAs	Indium arsenide
	InSb	Indium antimonide
(c) II–VI	ZnO	Zinc oxide
	ZnS	Zinc sulfide
	ZnSe	Zinc selenide
	ZnTe	Zinc telluride
	CdS	Cadmium sulfide
	CdSe	Cadmium selenide
	CdTe	Cadmium telluride
	HgS	Mercury sulfide
(d) IV–VI..........	PbS	Lead sulfide
	PbSe	Lead selenide
	PbTe	Lead telluride
(3) Alloys		
(a) Binary..........	$Si_{1-x}Ge_x$	
(b) Ternary........	$Al_xGa_{1-x}As$	
	$Al_xGa_{1-x}N$	
	$Al_xGa_{1-x}Sb$	
	$Cd_{1-x}Mn_xTe$	
	$GaAs_{1-x}P_x$	
	$Hg_{1-x}Cd_xTe$	
	$In_xAl_{1-x}As$	
	$In_xGa_{1-x}As$	
	$In_xGa_{1-x}N$	
(c) Quaternary..	$Al_xGa_{1-x}As_ySb_{1-y}$	
	$Ga_xIn_{1-x}As_{1-y}P_y$	

Although the number of semiconducting materials is reasonably large, the list is actually quite limited considering the total number of elements and possible combinations of elements. Note that, referring to the abbreviated periodic chart of the elements in Table 1.2, only a certain group of elements and elemental combinations typically give rise to semiconducting materials. Except for the IV–VI compounds, all of the

Table 1.2 Abbreviated Periodic Chart of the Elements

II	III	IV	V	VI
4 Be	5 B	6 C	7 N	8 O
12 Mg	13 **Al**	14 **Si**	15 **P**	16 S
30 **Zn**	31 **Ga**	32 Ge	33 **As**	34 **Se**
48 Cd	49 In	50 Sn	51 Sb	52 Te
80 Hg	81 Tl	82 Pb	83 Bi	84 Po

semiconductors listed in Table 1.1 are composed of elements appearing in column IV of the Periodic Table, or are a combination of elements in Periodic Table columns an equal distance from either side of column IV. The column III element Ga plus the column V element As yields the III–V compound semiconductor GaAs; the column II element Zn plus the column VI element Se yields the II–VI compound semiconductor ZnSe, the fractional combination of the column III elements Al and Ga plus the column V element As yields the alloy semiconductor $Al_xGa_{1-x}As$. This very general property is related to the chemical bonding in semiconductors, where, on the average, there are four valence electrons per atom.

Perhaps a word is in order concerning the obvious omissions from Table 1.1. Although researched as a semiconductor for high temperature applications, C (diamond), a column IV element, is typically classified as an insulator. Sn and Pb, also from column IV, are metals, although Sn is sometimes referred to as a semi-metal because it exhibits semiconductor-like properties at low temperatures. III–V compounds such as AlN and BN are fairly common insulators. Other possible compound combinations are not semiconductors, have not been thoroughly researched, exhibit undesirable physical properties, and/or seldom if ever appear in the device literature.

Another important general characteristic of the widely employed semiconductors is compositional purity. It is an established fact that even extremely minute traces of impurity atoms can have detrimental effects on the electrical properties of semiconductors. For this reason the compositional purity of semiconductors must be very carefully controlled: in fact, modern semiconductors are some of the purest solid materials in existence. The residual impurity concentrations listed in Table 1.3 are typical for the Si and GaAs employed in device fabrication. In examining this table it should be noted that there are approximately 5×10^{22} Si atoms per cm^3 and 2.2×10^{22} GaAs molecules per cm^3. Thus an impurity concentration of $5 \times 10^{16}/cm^3$ corresponds to about 1 impurity atom per 10^6 atoms of the parent semiconductor. The relatively high concentrations of oxygen and carbon in Si cited in Table 1.3 arise from the equipment (quartz crucible and graphite components) used in forming the Si single crystal. Oxygen actually

Table 1.3 Representative Impurity Concentrations in as-grown Si and GaAs Single Crystals.[†] (a) Derived from References [1] and [9]; (b) Derived from Reference [2].

(a) Si

Impurity	Concentration (10^{15} atoms/cm³)
O	500–1000
C	<25
Unintentional dopants (B, P, As, Sb)	≤0.05
Cr, Co, Ta, W, Na	Trace amounts (<0.005)
Cu, Au, Ni	Below detectable limits

(b) GaAs

Impurity	Concentration (10^{15} atoms/cm³)
C	0.1–1.0
Si	0.1
S	0.2
Se	0.4
Te	<0.1
Mn	<0.02
Be	<0.1
Mg	<0.03
B	4–50[‡]

[†] The concentrations quoted here are for Si crystals formed by the Czochralski-pulled method and GaAs crystals formed by the Liquid Encapsulated Czochralski (LEC) technique. The reader should be cautioned that impurity concentrations considerably above the cited values can be introduced during the subsequent handling and processing of semiconductor crystals.[1]

[‡] Pyrolithic boron nitride (PBN) crucibles are commonly used to hold the GaAs melt during crystal growth.

serves a useful role in that it increases the mechanical strength of Si, making the Si more robust in a manufacturing environment. Detrimental effects related to oxygen and carbon are minimized by proper device processing. The discussion here has of course been concerned with unintentional, mostly undesired impurities. Typically, dopant atoms at concentrations ranging from 10^{14}/cm³ to 10^{20}/cm³ are *purposely* added to a semiconductor to control its electrical properties.

The last material property we wish to address relates to the general spatial arrangement of atoms within device-quality semiconductors. The atomic arrangement within any given solid can be placed into one of three broad classifications: namely, amorphous, polycrystalline, or crystalline. An amorphous solid is a material in which there is no recognizable long-range order in the positioning of atoms within the material. The atomic arrangement in any given section of an amorphous material will look different from the atomic arrangement in any other section of the material. Crystalline solids lie at the opposite end of the "order" spectrum; in a crystalline material the atoms are arranged in an orderly three-dimensional array. Given any section of a crystalline material, one can readily reproduce the atomic arrangement in any other section of the material. Polycrystalline solids comprise an intermediate case in which the solid is composed of crystalline subsections that are disjointed or misoriented relative to each other.

Upon examining the many solid-state devices in existence, one finds examples of all three structural forms. An amorphous-Si thin-film transistor is used as the switching element in liquid crystal displays. Polycrystalline Si gates are employed in Metal-Oxide-Semiconductor Field-Effect Transistors (MOSFET's). In the vast majority of devices, however, the active region of the device is situated within a crystalline semiconductor. The overwhelming number of devices fabricated today employ crystalline semiconductors.

1.2 CRYSTAL STRUCTURE

Since device-quality semiconductors are typically crystalline in form, it is clearly desirable to accumulate additional information about the crystalline state. Our major goal here is to present a more detailed picture of the atomic arrangements within the various semiconductors. The use of unit cells to characterize the spatial positioning of atoms within crystals is first reviewed and then applied to simple three-dimensional lattices (atomic arrangements). The complete set of Bravais lattices and the division of solids into crystal systems are next discussed prior to examining semiconductor lattices themselves. The final two subsections are devoted to the introduction and use of Miller indices. Miller indices are a convenient shorthand notation widely employed for identifying specific planes and directions within crystals.

1.2.1 The Unit Cell Concept

Simply stated, a unit cell is a small portion of any given crystal that can be used to reproduce the crystal. To help establish the unit cell (or building-block) concept, let us consider the two-dimensional lattice shown in Fig. 1.1(a). In order to describe this lattice or to totally specify the physical characteristics of this lattice, one need only provide the unit cell shown in Fig. 1.1(b). As indicated in Fig. 1.1(c), the original lattice can be readily reproduced by merely duplicating the unit cell and stacking the duplicates next to each other in an orderly fashion.

The relationship between a given unit cell and the lattice it characterizes can be more precisely described in terms of *basis vectors*. If **a** is a vector of length a parallel to the a-side of the unit cell, and **b** is a vector of length b parallel to the b-side of the unit cell (see Fig. 1.1(d)), then equivalent points of a two-dimensional lattice will be separated by

$$\mathbf{r} = h\mathbf{a} + k\mathbf{b} \tag{1.1}$$

where h and k are integers. Hence, the lattice can be constructed by duplicating the unit cell and translating the duplicates $\mathbf{r} = \mathbf{a}$, $\mathbf{r} = \mathbf{b}$, $\mathbf{r} = \mathbf{a} + \mathbf{b}$, etc., relative to the original.

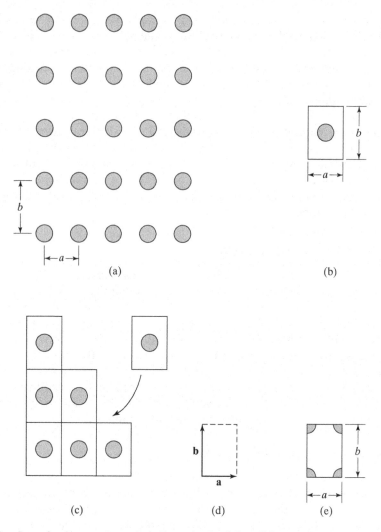

Figure 1.1   Introduction to the unit cell method of describing atom arrangements within crystals. (a) Sample two-dimensional lattice. (b) Unit cell corresponding to the part (a) lattice. (c) Reproduction of the original lattice. (d) Basis vectors. (e) An alternative unit cell.

In dealing with unit cells there often arises a misunderstanding, and hence confusion, relative to two points. First of all, unit cells are not necessarily unique. The unit cell shown in Fig. 1.1(e) is as acceptable as the Fig. 1.1(b) unit cell for specifying the original lattice of Fig. 1.1(a). Second, a unit cell need not be primitive (the smallest unit cell possible). In fact, it is usually advantageous to employ a slightly larger unit cell with orthogonal sides instead of a primitive cell with nonorthogonal sides. This is especially true in three dimensions where noncubic unit cells are quite difficult to describe and visualize.

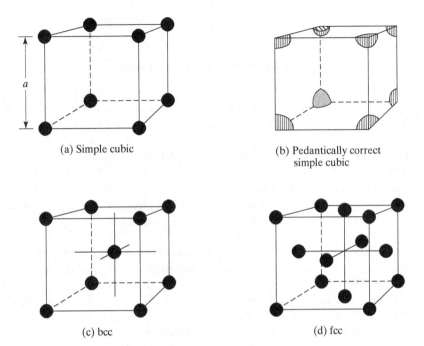

(a) Simple cubic

(b) Pedantically correct
simple cubic

(c) bcc

(d) fcc

Figure 1.2   Simple three-dimensional unit cells. (a) Simple cubic unit cell. (b) Pedantically correct simple cubic unit cell including only the fractional portion (1/8) of each corner atom actually within the cell cube. (c) Body-centered cubic unit cell. (d) Face-centered cubic unit cell (After Pierret.[3])

1.2.2   Simple 3-D Unit Cells

Semiconductor crystals are three-dimensional and are therefore described in terms of three-dimensional (3-D) unit cells. In Fig. 1.2(a) we have pictured the simplest of all 3-D unit cells—namely, the simple cubic unit cell. The simple cubic cell is an equal-sided box or cube with an atom positioned at each corner of the cube. The simple cubic lattice associated with this cell is constructed in a manner paralleling the two-dimensional case. In doing so, however, it should be noted that only 1/8 of each corner atom is actually *inside* the cell, as pictured in Fig. 1.2(b). Duplicating the Fig. 1.2(b) cell and stacking the duplicates like blocks in a nursery yields the simple cubic lattice. Alternatively, one could of course construct the lattice using the translation vectors $\mathbf{r} = h\mathbf{a} + k\mathbf{b} + l\mathbf{c}$, where $\mathbf{a}, \mathbf{b}$, and $\mathbf{c}$ are basis vectors and h, k, and l are integers.

Figures 1.2(c) and 1.2(d) display two common 3-D unit cells that are somewhat more complex but still closely related to the simple cubic cell. The unit cell of Fig. 1.2(c) has an atom added at the center of the cube; this configuration is appropriately called the body-centered cubic (bcc) unit cell. The face-centered cubic (fcc) unit cell of Fig. 1.2(d) contains an atom at each face of the cube in addition to the atoms at each corner. (Note, however, that only one-half of each face atom actually lies inside the fcc unit cell.) Whereas the simple cubic cell contains one atom (1/8 of an atom at each of the eight cube corners), the somewhat more complex bcc and fcc cells contain two and four atoms, respectively. The reader should verify these facts and visualize the lattices associated with the bcc and fcc cells.

1.2.3 Bravais Lattices and Crystal Systems

The sample 3-D unit cells considered in the preceding subsection are but three of many conceivable arrangements. The number of *unique point* lattices is, however, quite small. Bravais, in the 1840's, proved that there are just 14 different ways of arranging points in space latices such that all the lattice points have exactly the same surroundings. The 14 Bravais lattices (actually unit cells) are shown in Fig. 1.3 and of course include the simple cubic, bcc, and fcc cells. Upon examining Fig. 1.3 one might argue that there should be additional lattices, such as a face-centered tetragonal lattice. However, after a more detailed examination the proposed face-centered tetragonal and existing body-centered tetragonal lattices are found to be equivalent.

The points in Bravais lattices, it must be emphasized, are just that—points. There need not be a one-to-one correspondence between the atoms in a real crystal and points on a Bravais lattice. In fact, to characterize real crystal lattices it is often necessary to associate a group of atoms or a molecule with each point on a Bravais lattice. Naturally, this increases the number of unique lattice arrangements. When the geometrical properties of the molecules or groups of atoms at each lattice point are taken into account, one finds that there are 230 different repetitive patterns in which atomic elements can be arranged to form actual crystal structures. We point this out because the unit cells for semiconductor crystals are typically more complex than Bravais cells.

In crystallography, which deals with the cataloging and description of crystals, it is common practice to organize all crystals into *crystal systems*. The classification is based not on the Bravais lattice with which the crystal may be associated (as one might expect) but on the symmetry characteristics exhibited by the crystal. The symmetry properties used for classification are:

(1) *n-fold rotation symmetry*. Will the original crystal be reproduced if the crystal is rotated by an angle of $2\pi/n$ radians (n = 1, 2, 3, 4, 6) about an axis through the crystal?

(2) *Plane of symmetry*. Does there exist a plane in the crystal such that the lattice on one side of the plane is a mirror image of the lattice on the other side of the plane?

(3) *Inversion center symmetry*. Does there exist a point in the lattice such that the operation $\mathbf{r} \rightarrow -\mathbf{r}$ (where $\mathbf{r}$ is a vector from the inversion point to any other lattice point) leaves the lattice unchanged?

(4) *Rotation-inversion symmetry*. Will the original lattice be reproduced if one rotates the crystal by an angle of $2\pi/n$ radians (n = 1, 2, 3, 4, 6) and then passes all lattice points through an inversion center on the rotation axis?

When divided according to their symmetry properties, as summarized in Table 1.4 and the associated Fig. 1.4, all real crystals (and the Bravais lattices) fall into one of seven groupings—the seven crystal systems. Of particular relevance to our discussion, all the semiconductors previously listed in Table 1.1 are members of either the cubic system or the hexagonal system.

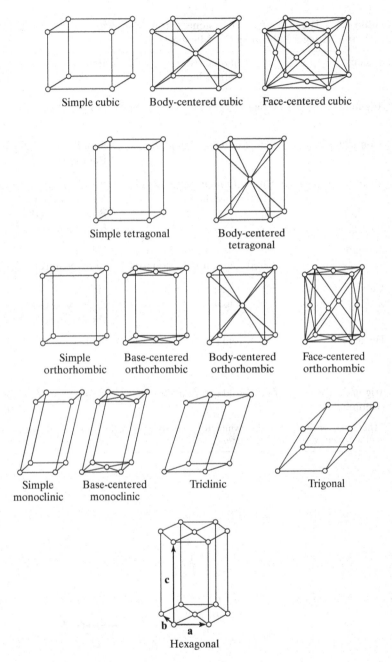

Figure 1.3   The 14 Bravais lattices. (From McKelvey.[4] Reproduced with permission, Robert E. Krieger Publishing Co., Malabar, FL.)

Table 1.4   The Seven Crystal Systems. (From McKelvey[4]. Reproduced with permission, Robert E. Krieger Publishing Co., Malabar, FL.)

System	Characteristic Symmetry Element[†]	Bravais Lattice	Unit Cell Characteristics
Triclinic	None	Simple	$a \neq b \neq c$ $\alpha \neq \beta \neq \gamma \neq 90°$
Monoclinic	One 2-fold rotation axis	Simple Base-centered	$a \neq b \neq c$ $\alpha = \beta = 90° \neq \gamma$
Orthorhombic	Three mutually perpendicular 2-fold rotation axes	Simple Base-centered Body-centered Face-centered	$a \neq b \neq c$ $\alpha = \beta = \gamma = 90°$
Tetragonal	One 4-fold rotation axis or a 4-fold rotation-inversion axis	Simple Body-centered	$a = b \neq c$ $\alpha = \beta = \gamma = 90°$
Cubic	Four 3-fold rotation axes (cube diagonals)	Simple Body-centered Face-centered	$a = b = c$ $\alpha = \beta = \gamma = 90°$
Hexagonal	One 6-fold rotation axis	Simple	$a = b \neq c$ $\alpha = 120°$ $\beta = \gamma = 90°$
Trigonal (Rhombohedral)	One 3-fold rotation axis	Simple	$a = b = c$ $\alpha = \beta = \gamma \neq 90°$

[†]There may, of course, be other symmetry properties in individual cases; only the one peculiar to each particular crystal system are listed here.

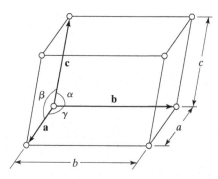

Figure 1.4   Definition of the angles and unit-cell dimensions cited in Table 1.4. (From McKelvey.[4] Reproduced with permission, Robert E. Krieger Publishing Co., Malabar, FL.)

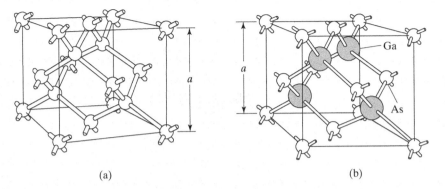

Figure 1.5   (a) Diamond lattice unit cell. (b) Zincblende lattice unit cell (GaAs used for illustration). [(a) After Shockley;[5] (b) after Sze.[6] Reprinted with permission.]

1.2.4 Specific Semiconductor Lattices

We are finally in a position to supply details relative to the positioning of atoms within semiconductor crystals. In Si and Ge the lattice structure is described by the unit cell pictured in Fig. 1.5(a). The Fig. 1.5(a) arrangement is known as the *diamond lattice* unit cell because it also characterizes diamond, a form of the column IV element carbon. Examining the diamond lattice unit cell, we note that the cell is cubic, with atoms at each corner and at each face of the cube, similar to the fcc cell. The interior of the Fig. 1.5(a) cell, however, contains four additional atoms. One of the interior atoms is located along a cube body diagonal exactly one-quarter of the way down the diagonal from the top front left-hand corner of the cube. The other three interior atoms are displaced one-quarter of the body diagonal length along the previously noted body diagonal direction from the front, top, and left-side face atoms, respectively. Although it may be difficult to visualize from Fig. 1.5(a), the diamond lattice can also be described as nothing more than two interpenetrating fcc lattices. The corner and face atoms of the unit cell can be viewed as belonging to one fcc lattice, while the atoms totally contained within the cell belong to the second fcc lattice. The second lattice is displaced one-quarter of a body diagonal along a body diagonal direction relative to the first fcc lattice.

Most of the III–V semiconductors, including GaAs, crystallize in the *zincblende* structure. The zincblende lattice, typified by the GaAs unit cell shown in Fig. 1.5(b), is essentially identical to the diamond lattice, except that lattice sites are apportioned equally between two different atoms. Ga occupies sites on one of the two interpenetrating fcc sublattices; arsenic (As) populates the other fcc sublattice.

Turning to the II–VI and IV–VI compound semiconductors we find greater structural variety. Some II–VI compounds crystallize in the zincblende lattice, others in the *wurtzite* lattice, while still others exhibit both zincblende and wurtzite structural forms. The IV–VI lead-based semiconductors, on the other hand, crystallize in the *rock-salt* lattice. Wurtzite (CdS) and rock-salt (PbS) unit cells are pictured in Fig. 1.6.

The diamond, zincblende, and rock-salt lattices all belong to the cubic crystal system and a single lattice constant *a* can be used to characterize the size of their unit cells.

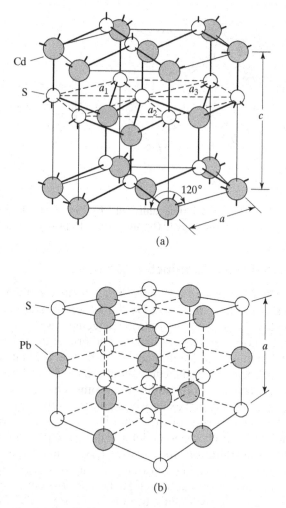

Figure 1.6   (a) Wurtzite lattice (CdS example). (b) Rock-salt lattice (PbS example). (From Sze.[6] Reproduced with permission.)

Table 1.5   Crystal Structure and 300 K Lattice Constants of Representative Materials (1 Å = 10^{-8} cm)

Semiconductor	Crystal Structure	Lattice Constant (Å)
Si	Diamond	5.43095
Ge	Diamond	5.64613
GaAs	Zincblende	5.6536
CdS	Zincblende	5.8320
	Wurtzite	$a = 4.16, c = 6.756$
PbS	Rock-Salt	5.9362

A summary of the crystal structure and the lattice constants for the semiconductors mentioned in this subsection are given in Table 1.5. (An expanded listing of semiconductor crystal structures and lattice parameters can be found in Appendix F of Sze[6].) From the data provided in Table 1.5, and the fact that there are 8 atoms per unit cell of volume a^3 in the diamond and zincblende lattices, one readily deduces, for example, a room-temperature atomic density of 4.99×10^{22} atoms/cm^3 for Si and a molecular density of 2.21×10^{22} molecules/cm^3 for GaAs. Exhibiting a unit cell with a six-sided base, the wurtzite lattice clearly belongs to the hexagonal crystal system. To specify the size of a hexagonal cell one must provide both the base side-length a and the height c of the unit cell. The hexagonal cell volume is $(3\sqrt{3}/2)a^2c$.

1.2.5 Miller Indices

A discussion of Miller indices, the accepted means for identifying planes and directions within a crystalline lattice, is a logical supplement to any crystal structure presentation. From a practical standpoint, a knowledge of Miller indices is often essential in dealing with semiconductor materials and device structures. In this subsection we cover the Miller indexing formalism; in the next subsection we examine sample practical applications of the formalism.

The Miller indices for any given plane of atoms within a crystal are obtained by following a straightforward four-step procedure. The procedure is detailed below, along with the simultaneous sample indexing of the plane shown in Fig. 1.7(a).

Indexing Procedure for Planes	Sample Implementation
(1) After setting up coordinate axes along the edges of the unit cell, note where the plane to be indexed intercepts the axes. Divide each intercept value by the unit cell length along the respective coordinate axis. Record the resulting normalized (pure-number) intercept set in the order x, y, z.	2, 1, 3
(2) Invert the intercept values—that is, form [1/intercept]s.	1/2, 1, 1/3
(3) Using an appropriate multiplier, convert the 1/intercept set to the smallest possible set of whole numbers.	3, 6, 2
(4) Enclose the whole-number set in curvilinear brackets.	(362)

To complete the description of the plane-indexing procedure, the user should also be aware of the following special facts:

  i) If the plane to be indexed is parallel to a coordinate axis, the "intercept" along that axis is taken to be at infinity. Thus for example, the plane shown in Fig. 1.7(b) intercepts the coordinate axes at $1, \infty, \infty$, and is therefore a (100) plane.

 ii) If the plane to be indexed has an intercept along the negative portion of a coordinate axis, a minus sign is placed *over* the corresponding index number. Thus the Fig. 1.7(c) plane is designated a $(1\bar{1}1)$ plane.

iii) All planes which fold into each other upon application of crystal-symmetry operations cannot be distinguished from each other by any physical measurement and

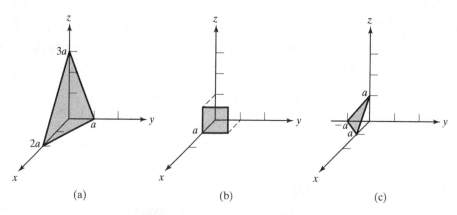

Figure 1.7 Sample cubic crystal planes. (a) The (362) plane used in explaining the Miller indexing procedure. (b) The (100) plane. (c) The ($1\bar{1}1$) plane.

are therefore said to be "equivalent." A group of equivalent planes is concisely referenced in the Miller notation through the use of braces, { }. In particular, according to Table 1.4 the cubic crystal system contains four three-fold rotation axes of symmetry passing through the cube diagonals. Examining the diamond lattice of Fig. 1.5(a), note that indeed exactly the same crystal structure is produced if one rotates the crystal 120° about any cube diagonal. When this symmetry operation is performed, all the cube faces fold into each other. In other words, the (100), (010), (001), ($\bar{1}$00), ($0\bar{1}0$), and ($00\bar{1}$) planes are equivalent planes and are collectively represented in Miller notation as {100} planes.

iv) Miller indices cannot be established for a plane passing through the origin of coordinates. The origin of coordinates must be moved to a lattice point not on the plane to be indexed. This procedure is acceptable because of the equivalent nature of parallel planes.

The Miller indices for *directions* are established in a manner analogous to the well-known procedure for finding the components of a vector. First, set up a vector of arbitrary length in the direction of interest. Next, decompose the vector into its basis vector (**a**, **b**, **c**) components by noting the projections of the direction vector along the coordinate axes. Using an appropriate multiplier, convert the coefficients of the basis vectors into the smallest possible whole-number set. This of course changes the length of the original vector but not its direction. Finally, with the direction of interest in the crystal specified by the vector $h\mathbf{a} + k\mathbf{b} + l\mathbf{c}$, where h, k, and l are positive or negative integers, the Miller notation for the direction becomes [hkl]. Note that square brackets, [ ], are used in the Miller notation to designate directions within a crystal; triangular brackets, ⟨ ⟩, designate an equivalent set of directions. Sample direction vectors and their corresponding Miller indices are displayed in Fig. 1.8. A summary of the Miller bracketing convention for planes and directions is given in Table 1.6.

In the foregoing discussion we presented the procedure for progressing from a given plane or direction in a crystal to the corresponding Miller indices. More often than not, one is faced with the inverse process—visualizing the crystalline plane or direction corresponding to a given set of indices. Fortunately, one seldom encounters

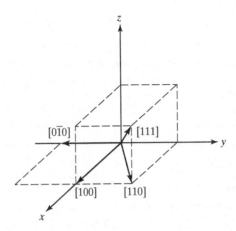

Figure 1.8   Sample direction vectors and their corresponding Miller indices.

Table 1.6   Miller Convention Summary

Convention	Interpretation	
(hkl)	Crystal plane	
$\{hkl\}$	Equivalent planes	
$[hkl]$	Crystal direction	
$\langle hkl \rangle$	Equivalent directions	
$(hkil)$	Plane	Hexagonal
$[hkil]$	Direction	system

other than low-index planes and directions such as (111), (110), [001], etc. Thus it is possible to become fairly adept at the inverse process by simply memorizing the orientations of planes and directions associated with small-number indices. It is also helpful to note that, *for cubic crystals, a plane and the direction normal to the plane have precisely the same indices*—e.g., the [110] direction is normal to the (110) plane. Of course, any plane or direction can always be deduced by reversing the indexing procedure.

When dealing with planes and directions in a crystal, one is often specifically interested in the distance between parallel planes containing identical atomic arrangements or the angle between directions. For a cubic crystal with lattice constant a, the separation d between adjacent (hkl) planes is readily shown to be

$$d = \frac{a}{\sqrt{h^2 + k^2 + l^2}} \tag{1.2}$$

Likewise for a cubic crystal, the angle θ between directions $[h_1 k_1 l_1]$ and $[h_2 k_2 l_2]$ is given by

$$\cos(\theta) = \frac{h_1 h_2 + k_1 k_2 + l_1 l_2}{[(h_1^2 + k_1^2 + l_1^2)(h_2^2 + k_2^2 + l_2^2)]^{1/2}} \tag{1.3}$$

Equation (1.3) is established by forming the dot product between the direction vectors $r_1 = h_1\mathbf{a} + k_1\mathbf{b} + l_1\mathbf{c}$ and $r_2 = h_2\mathbf{a} + k_2\mathbf{b} + l_2\mathbf{c}$. As one might expect, d and θ relationships valid for the other crystal systems are somewhat more complex.[7,8]

Since a number of semiconductors crystallize in the hexagonal wurtzite lattice, it is worthwhile to call attention to the special four-digit indices (called Miller-Bravais indices) commonly used for identifying planes and directions in hexagonal crystals. When working with hexagonal crystals, rather than employing orthogonal coordinates, it is more convenient to set up three non-orthogonal basis vectors $\mathbf{a}_1$, $\mathbf{a}_2$, and $\mathbf{a}_3$ in the base plane of the unit cell. The orientations of these vectors are pictured in Fig. 1.9 along with the height basis vector $\mathbf{c}$. Indices for planes and directions are formed in a manner paralleling the previously described three-digit indexing scheme. The resulting designations are of the forms $(hkil)$ and $[hkil]$. For example, (0001) is a plane parallel to the base of the unit cell. $(10\bar{1}0)$ describes the face plane on the side of the unit cell having intercepts of $1, \infty, -1$, and ∞ on the a_1, a_2, a_3, and c axes, respectively. Please note that, because of the chosen orientations of $\mathbf{a}_1$, $\mathbf{a}_2$, and $\mathbf{a}_3$, the first three Miller-Bravais indices must always sum to zero—i.e., $h + k + i = 0$.

1.2.6 Example Use of Miller Indices

A knowledge and understanding of Miller indices are all but essential in device-related work, particularly device fabrication. To promote assimilation of the formalism, we consider three practical applications. The chosen examples also provide supplemental information of a generally useful and relevant nature.

Wafer Surface Orientation

The fabrication of Si devices typically begins with a single crystal of Si in the form of a thin circular "wafer." Wafers can be purchased in standard sizes ranging from approximately 1 inch to 12 inches in diameter. Cut from larger, cylindrically-shaped, single

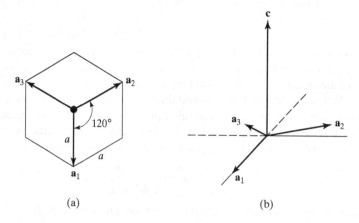

(a)   (b)

Figure 1.9 Basis vectors for hexagonal crystals. (a) Base-plane orientation. (b) Three-dimensional orientation.

crystals called ingots, wafers are carefully polished, etched, and shaped prior to being purchased. To maintain reasonable device yields and to assure compatibility with processing equipment, the wafers must meet very stringent specifications. For example, the largest wafers presently employed in the manufacture of ICs are specified to have a diameter of 300 ± 0.2 mm, a thickness of 775 ± 25 μm, and a (100) ± 0.2° surface orientation. In general, the surface orientation of standard Si wafers is either a (100) or (111) plane.[†] Circular wafers are readily produced with the cited surface orientations because the axial growth direction of cylindrically shaped Si ingots is either [100] or [111].

Wafer Flats and Notches

Wafer flats are straight-line regions along the periphery of a wafer as pictured in Fig. 1.10. The longer primary flat is positioned to facilitate identification of crystalline directions lying within the surface plane. The positioning of the shorter secondary flat relative to the primary flat indicates by inspection the wafer type (*n* or *p*) and the surface

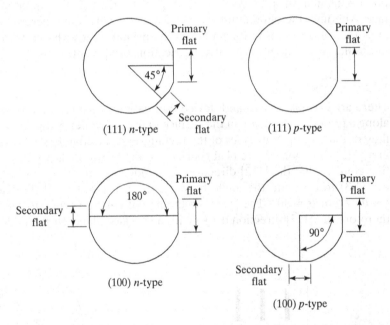

Figure 1.10   Convention for identification of flats on silicon wafers. (Reprinted with permission, from the Semiconductor Equipment and Materials Institute, Inc., Book of SEMI Standards. Copyright the Semiconductor Equipment and Materials Institute, Inc., 625 Ellis St., Suite 212, Mountain View, CA 94043.)

[†]To be precise, the surface is a {100} or {111} plane. When referring to surface planes, however, it is common practice to use the characteristic member of the equivalent set—e.g., (100) of the {100} set of planes.

orientation [(100) or (111)]. For wafers with a (100) surface plane, the primary flat along the edge of the wafer is a (011) plane, and the normal to the flat lying within the surface plane is a [011] direction. (If the surface of the wafer is considered to be a (001) plane, the primary flat would be a (110) plane and the normal a [110] direction. It is sometimes more convenient to make this alternative identification.) For (111) wafers, the normal to the primary flat lies in the [$\bar{1}$10] direction. Identification of the noted directions in the (100) and (111) surface planes allows the positioning of devices to achieve the maximum yield when dividing the wafer into segments (called die) containing individual devices. With the identification of one direction in the surface plane, it also becomes possible to deduce the orientation of any other direction in the surface plane. This is useful when optimal operating conditions require the special positioning of a pattern or device on the surface of the wafer.

It should be noted that wafer flats are routinely found on wafers up to 6 inches in diameter. However, bigger diameter wafers purchased in large quantities with identical specifications often contain only a small notch, a semicircular 1 mm deep indentation, along the wafer periphery. The direction normal to the wafer periphery at the notch point is again a [011] direction on a (100) surface and a [$\bar{1}$10] direction on a (111) surface. Although the elimination of wafer flats permits some increase in usable surface area, the primary reason for replacing the primary flat with a notch is to better facilitate wafer handling by automated fabrication equipment.

Pattern Alignment

There are certain devices and device fabrication processes that require an alignment along a preferred direction in the surface plane of a wafer. Suppose for the purposes of illustration that the long sides of the rectangles pictured in Fig. 1.11(a) must be aligned in a [11$\bar{2}$] direction on the (111) surface of the Si wafer shown in Fig. 1.11(b). Let us first confirm that the [11$\bar{2}$] direction does indeed lie in the (111) surface plane. Using Eq. (1.3) to compute the angle between the [111] and [11$\bar{2}$] directions, one obtains $\cos \theta = 0$ or $\theta = 90°$. The [111] direction is of course normal to the (111) plane, and therefore the [11$\bar{2}$] direction does lie in the surface plane. Next, employing Eq. (1.3) to

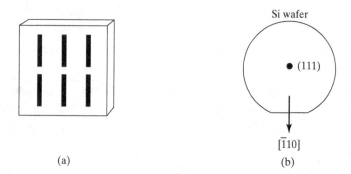

(a)   (b)

Figure 1.11   Pattern alignment illustration. (a) Pattern to be aligned. (b) *p*-type Si wafer with a (111) surface orientation.

compute the angle between the [$\bar{1}$10] normal to the primary flat and the [11$\bar{2}$] direction, one again obtains $\cos\theta = 0$ or $\theta = 90°$. Consequently, the desired alignment is achieved by rotating the pattern 90° relative to the flat normal—i.e., the long sides of the rectanges are to be aligned parallel to the edge of the primary flat.

REFERENCES

[1] P. F. Schmidt and C. W. Pearce, "A Neutron Activation Analysis of the Sources of Transition Group Metal Contamination," J. Electrochem. Soc., *128*, 630 (March, 1981).

[2] M. R. Brozel, "Defect Densities in Melt-Grown GaAs (a Review)," in *Properties of Gallium Arsenide*, 3rd edition, edited by M. R. Brozel and G. E. Stillman, INSPEC, London, 1996.

[3] R. F. Pierret, *Semiconductor Device Fundamentals*, Addison-Wesley, Reading, MA, 1996.

[4] J. P. McKelvey, *Solid-State and Semiconductor Physics*, Harper and Row, New York, 1966.

[5] W. Shockley, *Electrons and Holes in Semiconductors*, Litton Educational Publishing, Inc., 1950.

[6] S. M. Sze, *Physics of Semiconductor Devices*, 2nd edition, John Wiley & Sons, New York, 1981.

[7] W. R. Runyan and T. J. Shaffner, *Semiconductor Measurements and Instrumentation*, 2nd edition, McGraw-Hill, New York, 1998. (See Table 1.8 on p. 26.)

[8] F. C. Phillips, *An Introduction to Crystallography*, Longmans, Green and Co., New York, 1957. (See p. 202.)

[9] J. D. Plummer, M. D. Deal, and P. B. Griffin, *Silicon VLSI Technology, Fundamentals, Practice and Modeling*, Prentice Hall, Upper Saddle River, NJ, 2000.

SOURCE LISTING

References [1] through [4], [6], [7], and [9] are all good sources of additional relevant information.

PROBLEMS

1.1 A 3-D cubic unit cell contains four atoms whose centers are positioned halfway up along each of the vertical cell edges and one atom each centered in the middle of the top and bottom faces.
  (a) How many atoms are there per unit cell (atoms actually inside the unit cube)?
  (b) Make a sketch (as best you can) of the lattice characterized by the part (a) unit cell. Use dots to represent the atoms and include at least three parallel planes of atoms.
  (c) What is the name of the *standard* unit cell used to characterize the lattice sketched in part (b)?

1.2 (a) How many atoms are there in the unit cell characterizing the Si lattice?
  (b) Verify that there are 4.994×10^{22} atoms/cm^3 in the Si lattice at room temperature.
  (c) Determine the center-to-center distance between nearest neighbors in the Si lattice.

1.3 AlN crystallizes in the wurtzite lattice with $a = 3.1115$ Å and $c = 4.9798$ Å at 300 K. Determine the number of nitrogen atoms per cm^3 in the AlN crystal at 300 K.

1.4 Record all intermediate steps in answering the following questions.
   (a) As shown in Fig. P1.4(a), a crystalline plane has intercepts of $6a$, $3a$, and $2a$ on the x, y, and z axes, respectively. a is the cubic cell side length.
      (i) What is the Miller index notation for the plane?
      (ii) What is the Miller index notation for the direction normal to the plane?
   (b) Determine the Miller indices for the cubic crystal plane pictured in Fig. P1.4(b).
   (c) Given a hexagonal crystal structure, determine the Miller-Bravais indices for the plane pictured in Fig. P1.4(c). (The intercepts are one unit cell length along the a_1, a_2 and c axes.)

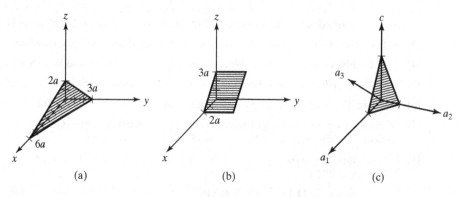

(a)   (b)   (c)

Figure P1.4

1.5 Referring to the unit cell of the Si lattice reproduced in Fig. P1.5, and noting that the origin of coordinates is located at the lower back corner of the unit cell:
   (a) What are the Miller indices of the plane passing through the points ABC?
   (b) What are the Miller indices of the plane passing through the points BCD?
   (c) What are the Miller indices of the direction vector running from the origin of coordinates to the point D?
   (d) What are the Miller indices of the direction vector running from the origin of coordinates to the point E?

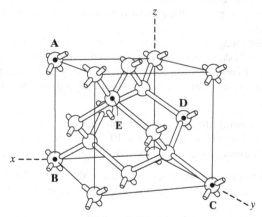

Figure P1.5

1.6 Conceptually position the origin of coordinates of an x-y-z axes system at the far back corner of the Fig. 1.5(b) unit cell and run the coordinate axes along the edges of the cell (z-upward). For each of the following planes, (i) sketch the orientation of the plane relative to the unit cell and (ii) indicate the arrangement of Ga and As atoms on the plane.
  (a) (001)
  (b) (100)
  (c) (011)
  (d) (111)

1.7 (a) The surface of a silicon wafer is a (100) plane. Which, if any, ⟨110⟩ directions lie in the (100) surface plane of the wafer?
  (b) Identify two crystalline directions that are perpendicular to the [111] direction in a cubic crystal. (NOTE: [$\bar{1}$11], [1$\bar{1}$1], etc., are NOT perpendicular to the [111] direction.)

1.8 Consider the PbS unit cell pictured in Fig. 1.6(b). The lattice constant for PbS is $a = 5.9362$ Å.
  (a) To which crystal system does PbS belong?
  (b) Determine the number of Pb atoms/cm^3 in the PbS lattice.
  (c) Suppose the origin of coordinates of an x-y-z axes system is located at the lower back corner of the PbS cell and the coordinate axes are run along the edges of the cell (z upward). Determine the number of Pb atoms/cm^2 on a (120) plane. Record all your work.

1.9 A Si wafer will tend to cleave (break apart) along {111} planes if sufficient stress is applied to the surface of the wafer. If the top surface of the wafer is a (100) plane as pictured in Fig. P1.9:

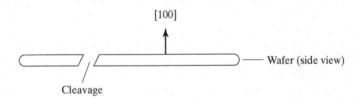

[100]

Wafer (side view)

Cleavage

Figure P1.9

  (a) What are the possible angles between the normal to the top surface and the cleavage planes?
  (b) If pressure is applied to a point on the surface of the wafer and {111} plane cleavage occurs through the pressure point, into how many pieces at maximum will the wafer break? (A million pieces is not the correct answer.)
  (c) Assuming cleavage occurs along a {111} plane, how will the broken edge of the wafer be oriented relative to the primary wafer flat?

1.10 A cylindrical Si ingot is produced whose axis is oriented in the [001] direction. A flat is subsequently machined along the side of the cylinder forming a (110) plane as shown in Fig. P1.10.
A research program requires wafers whose surfaces are (112) planes. Indicate how the ingot
must be sawed to achieve the desired wafers. Record your reasoning.

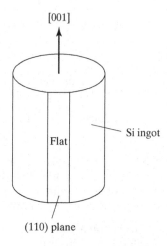

Figure P1.10

CHAPTER 2

Elements of Quantum Mechanics

In Chapter 1 we examined the physical nature of semiconductors and emphasized their typically crystalline structure. With an overriding interest in the electrical properties of semiconductors, our ultimate goal is of course to describe or model the action of electrons in these crystalline solids. Unfortunately, the "everyday" descriptive formalism known as classical (Newtonian) mechanics is found to be inaccurate when applied to electrons in crystals or, more generally, when applied to any system with atomic dimensions. To model the action of electrons in crystals one must employ the more involved mathematical formalism known as *Quantum Mechanics*. Quantum mechanics is a more precise description of nature which reduces to classical mechanics in the limit where the masses and energies of the particles involved are large.

Essential elements of the quantum mechanical formalism are presented and examined in this chapter. Our purpose is to provide the necessary background knowledge for understanding the subsequent treatment of electrons in crystals. Historically important experimental observations dealing with blackbody radiation, optical spectra emitted by atoms, and the wave-like nature of particles are first discussed to exhibit the failure of classical mechanics, and the success of quantum mechanics, in describing the behavior of systems with atomic dimensions. We next review basic mathematical aspects of the quantum mechanical formalism. Finally, simple problem solutions are considered to illustrate use of the formalism and to establish an information base for future reference.

2.1 THE QUANTUM CONCEPT

2.1.1 Blackbody Radiation

It is a well-known fact that a solid object will glow or give off light if it is heated to a sufficiently high temperature. Actually, solid bodies in equilibrium with their surroundings emit a spectrum of radiation at all times. When the temperature of the body is at or below room temperature, however, the radiation is almost exclusively in the infrared

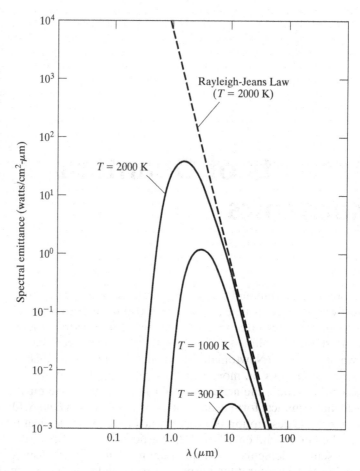

Figure 2.1   Wavelength dependence of the radiation emitted by a blackbody heated to 300 K, 1000 K, and 2000 K. Note that the visible portion of the spectrum is confined to wavelengths $0.4\,\mu m \lesssim \lambda \lesssim 0.7\,\mu m$. The dashed line is the predicted dependence for $T = 2000$ K based on classical considerations.

and therefore not detectable by the human eye. For an ideal radiator, called a blackbody, the spectrum or wavelength dependence of the emitted radiation is as graphed in Fig. 2.1.

Various attempts to explain the observed blackbody spectrum were made in the latter half of the 19th century. The most successful of the arguments, all of which were based on classical mechanics, was proposed by Rayleigh and Jeans. Heat energy absorbed by a material was known to cause a vibration of the atoms within the solid. The vibrating atoms were modeled as harmonic oscillators with a spectrum of normal mode frequencies, $\nu = \omega/2\pi$, and a *continuum of allowed energies* distributed in accordance with statistical considerations. The emitted radiation was in essence equated to a sampling of the energy distribution inside the solid. The Rayleigh–Jeans "law" resulting from this analysis is shown as a dashed line in Fig. 2.1. As is evident from Fig. 2.1, the classical theory was in reasonably good agreement with experimental observations at the longer wavelengths. Over the short-wavelength portion of the spectrum, however, there was total divergence between experiment and theory. This came to be known as the "ultraviolet catastrophe," since integration over all wavelengths theoretically predicted an infinite amount of radiated energy.

In 1901 Max Planck provided a detailed theoretical fit to the observed blackbody spectrum. The explanation was based on the then-startling hypothesis that the vibrating atoms in a material could only radiate or absorb energy in discrete packets. Specifically, for a given atomic oscillator vibrating at a frequency v, Planck postulated that the energy of the oscillator was restricted to the *quantized* values

$$E_{\mathbf{n}} = \mathbf{n} h v = \mathbf{n} \hbar \omega \qquad \mathbf{n} = 0, 1, 2, \cdots \qquad (2.1)$$

An h value of 6.63×10^{-34} joule-sec ($\hbar = h/2\pi$) was obtained by matching theory to experiment and has subsequently come to be known as Planck's constant.

The point to be learned from the blackbody discussion is that, for atomic dimension systems, the classical view, which always allows a continuum of energies, is demonstrably incorrect. Extremely small discrete steps in energy, or energy quantization, can occur and is a central feature of quantum mechanics.

2.1.2 The Bohr Atom

Another experimental observation which puzzled scientists of the 19th century was the sharp, discrete spectral lines emitted by heated gases. The first step toward unraveling this puzzle was provided by Rutherford, who advanced the nuclear model for the atom in 1910. Atoms were viewed as being composed of electrons with a small rest mass m_0 and charge $-q$ orbiting a massive nucleus with charge $+Zq$, where Z was an integer equal to the number of orbiting electrons. Light emission from heated atoms could then be associated with the energy lost by electrons in going from a higher-energy to a lower-energy orbit. Classically, however, the electrons could assume a continuum of energies and the output spectrum should likewise be continuous—not sharp, discrete spectral lines. The nuclear model itself posed somewhat of a dilemma. According to classical theory, whenever a charged particle is accelerated, the particle will radiate energy. Thus, based on classical arguments, the angularly accelerated electrons in an atom should continuously lose energy and spiral into the nucleus in a relatively short period of time.

In 1913 Niels Bohr proposed a model that both resolved the Rutherford atom dilemma and explained the discrete nature of the spectra emitted by heated gases. Building on Planck's hypothesis, Bohr suggested that the electrons in an atom were restricted to certain well-defined orbits, or, equivalently, assumed that the orbiting electrons could take on only certain (quantized) values of angular momentum L.

For the simple hydrogen atom with $Z = 1$ and a circular electron orbit, the Bohr postulate can be expressed mathematically in the following manner:

$$L_{\mathbf{n}} = m_0 v r_{\mathbf{n}} = \mathbf{n} \hbar \qquad \mathbf{n} = 1, 2, 3, \cdots \qquad (2.2)$$

where m_0 is the electron rest mass, v is the linear electron velocity, and r_n is the radius of the orbit for a given value of $\mathbf{n}$. Since the electron orbits are assumed to be stable, the centripedal force on the electron $(m_0 v^2 / r_n)$ must precisely balance the coulombic attraction $(q^2 / 4\pi\varepsilon_0 r_n^2$ in rationalized MKS units) between the nucleus and the orbiting electron. Therefore, one can also write

$$\frac{m_0 v^2}{r_n} = \frac{q^2}{4\pi\varepsilon_0 r_n^{\,2}} \tag{2.3}$$

where ε_0 is the permittivity of free space. Combining Eqs. (2.2) and (2.3), one obtains

$$r_n = \frac{4\pi\varepsilon_0 (\mathbf{n}\hbar)^2}{m_0 q^2} \tag{2.4}$$

Next, by examining the kinetic energy (K.E.) and potential energy (P.E.) components of the total electron energy (E_n) in the various orbits, we find

$$\text{K.E.} = \frac{1}{2} m_0 v^2 = \frac{1}{2}\, (q^2 / 4\pi\varepsilon_0 r_n) \tag{2.5a}$$

and

$$\text{P.E.} = -q^2 / 4\pi\varepsilon_0 r_n \quad (\text{P.E. set} = 0 \text{ at } r = \infty) \tag{2.5b}$$

Thus

$$E_n = \text{K.E.} + \text{P.E.} = -\frac{1}{2}\, (q^2 / 4\pi\varepsilon_0 r_n) \tag{2.6}$$

or, making use of Eq. (2.4),

$$\boxed{E_n = -\frac{m_0 q^4}{2(4\pi\varepsilon_0 \mathbf{n}\hbar)^2} = -\frac{13.6}{\mathbf{n}^2}\,\text{eV}} \tag{2.7}$$

The *electron volt* (eV) introduced in Eq. (2.7) is a non-MKS unit of energy equal to 1.60×10^{-19} joules.

With the electron energies in the hydrogen atom restricted to the values specified by Eq. (2.7), the light energies that can be emitted by the atom upon heating are now discrete in nature and equal to $E_{n'} - E_n$, $\mathbf{n}' > \mathbf{n}$. As summarized in Fig. 2.2, the allowed energy transitions are found to be in excellent agreement with the observed photo-energies.

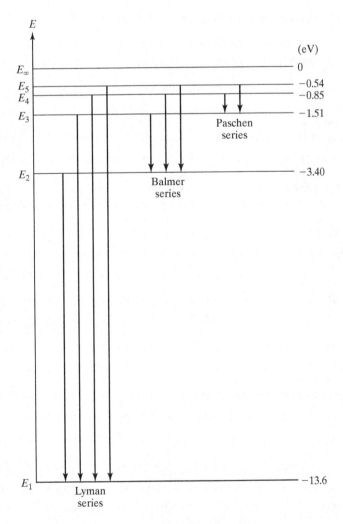

Figure 2.2 Hydrogen atom energy levels as predicted by the Bohr theory and the transitions corresponding to prominent, experimentally observed, spectral lines.

Although the Bohr model was immensely successful in explaining the hydrogen spectra, numerous attempts to extend the "semi-classical" Bohr analysis to more complex atoms such as helium proved to be futile. Success along those lines had to await further development of the quantum mechanical formalism. Nevertheless, the Bohr analysis reinforced the concept of energy quantization and the attendant failure of classical mechanics in dealing with systems on an atomic scale. Moreover, the quantization of angular momentum in the Bohr model clearly extended the quantum concept, seemingly suggesting a general quantization of atomic-scale observables.

2.1.3   Wave-Particle Duality

An interplay between light and matter was clearly evident in the blackbody and Bohr atom discussions. Those topics can be treated, however, without disturbing the classical viewpoint that electromagnetic radiation (light, X-rays, etc.) is totally wave-like in nature and matter (an atom, an electron) is totally particle-like in nature. A different situation arises in treating the photoelectric effect—the emission of electrons from the illuminated surface of a material. To explain the photoelectric effect, as argued by Einstein in 1905, one must view the impinging light to be composed of particle-like quanta (photons) with an energy $E = h\nu$. The particle-like properties of electromagnetic radiation were later solidified in the explanation of the Compton effect. The deflected portion of an X-ray beam directed at solids was found to undergo a change in frequency. The observed change in frequency was precisely what one would expect from a "billiard ball" type collision between the X-ray quanta and electrons in the solid. In such a collision both energy and momentum must be conserved. Noting that $E = h\nu = mc^2$, where m is the "mass" of the photon and c the velocity of light, the momentum of the photon was taken to be $p = mc = h\nu/c = h/\lambda$, λ being the wavelength of the electromagnetic radiation.

By the mid-1920's the wave-particle duality of electromagnetic radiation was an established fact. Noting this fact and the general reciprocity of physical laws, Louis de Broglie in 1925 made a rather interesting conjecture. He suggested that since electromagnetic radiation exhibited particle-like properties, particles should be expected to exhibit wave-like properties. De Broglie further hypothesized that, paralleling the photon momentum calculation, the *wavelength* characteristic of a given particle with momentum p could be computed from

$$\boxed{p = h/\lambda} \qquad \ldots \text{de Broglie hypothesis} \qquad (2.8)$$

Although pure conjecture at the time, the de Broglie hypothesis was quickly substantiated. Evidence of the wave-like properties of matter was first obtained by Davisson and Germer from an experiment performed in 1927. In their experiment, a low-energy beam of electrons was directed perpendicularly at the surface of a nickel crystal. The energy of the electrons was chosen such that the wavelength of the electrons as computed from the de Broglie relationship was comparable to the nearest-neighbor distance between nickel atoms. If the electrons behaved as simple particles, one would expect the electrons to scatter more or less randomly in all directions from the surface of the nickel crystal (assumed to be rough on an atomic scale). The angular distribution actually observed was quite similar to the interference pattern produced by light diffracted from a grating. In fact, the angular positions of maxima and minima of electron intensity could be predicted accurately using the de Broglie wavelength and assuming wave-like reflection from atomic planes inside the nickel crystal. Later experiments performed by other researchers likewise confirmed the inherent wave-like properties of heavier particles such as protons and neutrons.

In summary, then, based on experimental evidence—a portion of which has been discussed herein under the headings of blackbody radiation, the Bohr atom, and the wave-particle duality—one is led to conclude that classical mechanics does not accurately describe the action of particles on an atomic scale. Experiments point to a quantization of observables (energy, angular momentum, etc.) and to the inherent wave-like nature of all matter.

2.2 BASIC FORMALISM

2.2.1 General Formulation

The accumulation of experimental data and physical explanations in the early 20th century that were at odds with the classical laws of physics emphasized the need for a revised formulation of mechanics. In 1926 Schrödinger not only provided the required revision, but established a unified scheme valid for describing both the microscopic and macroscopic universes. The formulation, called *wave mechanics*, incorporated the physical notions of quantization first advanced by Planck and the wave-like nature of matter hypothesized by de Broglie. It should be mentioned that at almost the same time an alternative formulation called *matrix mechanics* was advanced by Heisenberg. Although very different in their mathematical orientations, the two formulations were later shown to be precisely equivalent and were merged under the general heading of *quantum mechanics*. Herein we will restrict ourselves to the Schrödinger wave mechanical description, which is somewhat simpler mathematically and more readily related to the physics of a particular problem. Nevertheless, the reader should be forewarned that problem-solving using wave mechanics is considerably different, and typically more involved, than a classical analysis. Our general approach will be to present the five basic postulates of wave mechanics and to subsequently discuss the postulates to provide some insight into the formulation.

For a single-particle system, the five basic postulates of wave mechanics are as follows:

(1) There exists a wavefunction, $\Psi = \Psi(x, y, z, t)$, from which one can ascertain the dynamic behavior of the system and all desired system variables. Ψ might be called the "describing function" for the system. Mathematically, Ψ is permitted to be a complex quantity (with real and imaginary parts) and will, in general, be a function of the space coordinates (x, y, z) and time t.

(2) The Ψ for a given system and specified system contraints is determined by solving the equation,

$$-\frac{\hbar^2}{2m}\nabla^2\Psi + U(x, y, z)\Psi = -\frac{\hbar}{i}\frac{\partial\Psi}{\partial t} \tag{2.9}$$

where m is the mass of the particle, U is the potential energy of the system,[†] and $i = \sqrt{-1}$. Equation (2.9) is referred to as the time-dependent Schrödinger equation, or simply, the wave equation.

[†] In analyses using classical mechanics one normally considers the force, $\mathbf{F}$, acting on a particle. Note that, since $\mathbf{F} = \nabla U$, forces indirectly enter the wave mechanics formulation through the potential energy U.

(3) Ψ and $\nabla\Psi$ must be finite, continuous, and single-valued for all values of x, y, z and t.

(4) If Ψ^* is the complex conjugate of Ψ, $\Psi^*\Psi d\mathcal{V} = |\Psi|^2 d\mathcal{V}$ is to be identified as the probability that the particle will be found in the spatial volume element $d\mathcal{V}$. Hence, by implication,

$$\int_{\mathcal{V}} \Psi^*\Psi d\mathcal{V} = 1 \qquad (2.10)$$

where $\int_{\mathcal{V}}$ indicates an integration over all space.

(5) One can associate a unique mathematical operator with each dynamic system variable such as position or momentum. The value—or, more precisely, the expectation value—of a given system variable is in turn obtained by "operating" on the wavefunction. Specifically, taking α to be the system variable of interest and α_{op} the associated mathematical operator, the desired expectation value, $\langle\alpha\rangle$, is computed from

$$\langle\alpha\rangle = \int_{\mathcal{V}} \Psi^*\alpha_{op}\Psi d\mathcal{V} \qquad (2.11)$$

The unique mathematical operator associated with a given system variable has been established by requiring the wave mechanical expectation value to approach the corresponding value derived from classical mechanics in the large-mass/high-energy limit. An abbreviated listing of dynamic variables and associated operators is presented in Table 2.1.

The solution of problems using wave mechanics is in principle quite straightforward. Subject to the constraints (boundary conditions) inherent in a problem and the additional constraints imposed by postulates 3 and 4, one solves Schrödinger's equation for the system wavefunction Ψ. Once Ψ is known, system variables of interest can be deduced from Eq. (2.11) per the postulate 5 recipe. The straightforward approach, however, is often difficult to implement. Except for simple problems of an idealized nature and a very select number of practical problems, it is usually impossible to obtain

Table 2.1 Dynamic Variable/Operator Correspondence

Dynamic Variable (α)		Mathematical Operator (α_{op})		Expectation Value—$\langle\alpha\rangle$
x, y, z	$\leftrightarrow$	x, y, z	$\cdots$	$\langle x\rangle = \displaystyle\int_{\mathcal{V}} \Psi^*x\Psi d\mathcal{V}$
$f(x, y, z)$	$\leftrightarrow$	$f(x, y, z)$		
p_x, p_y, p_z	$\leftrightarrow$	$\dfrac{\hbar}{i}\dfrac{\partial}{\partial x}, \dfrac{\hbar}{i}\dfrac{\partial}{\partial y}, \dfrac{\hbar}{i}\dfrac{\partial}{\partial z}$	$\cdots$	$\langle p_x\rangle = \displaystyle\int_{\mathcal{V}} \Psi^*\dfrac{\hbar}{i}\dfrac{\partial\Psi}{\partial x}d\mathcal{V}$
E	$\leftrightarrow$	$-\dfrac{\hbar}{i}\dfrac{\partial}{\partial t}$		

a closed-form solution to Schrödinger's equation. Nevertheless, in many problems, constraints imposed on the solution can be used to deduce information about the system variables, notably the allowed system energies, without actually solving for the system wavefunction. Another common approach is to use expansions, trial (approximate) wavefunctions, or limiting-case solutions to deduce information of interest.

Attention should also be drawn to a property of matter inherent in postulates 4 and 5 that is not apparent on a macroscopic scale. Namely, the exact location of a particle and its precise trajectory cannot be specified—one can only ascertain the *probability* of finding the particle in a given spatial volume and the *expectation* values of variables. Consistent with human perception, however, if quantum mechanics is applied to a massive object such as a baseball, $\Psi^*\Psi \, \mathrm{d}\mathcal{V}$ is found to be large only within the classical boundaries of the object, and the object is predicted to move in accordance with Newton's laws.

Finally, a comment is in order concerning the "derivation" of Schrödinger's equation and the origin of the other basic postulates. Although excellent theoretical arguments can be presented to justify the form of the equation,[1] Schrödinger's equation is essentially an empirical relationship. Like Newton's laws, Schrödinger's equation and the other basic postulates of quantum mechanics constitute a generalized mathematical description of the physical world extrapolated from specific empirical observations. Relative to the validity of the formulation, it can only be stated that, whenever subject to test by experiment, the predictions of the quantum mechanical formulation have been found to be in agreement with observations to within the limit of experimental uncertainty, which in many cases has been extremely small.[2]

2.2.2 Time-Independent Formulation

If the particle in the system under analysis has a fixed total energy E, the quantum mechanical formulation of the problem is significantly simplified. Consider the general expression for the energy expectation value as deduced from Eq. (2.11) and Table 2.1:

$$\langle E \rangle = \int_{\mathcal{V}} \Psi^* \left(-\frac{\hbar}{i} \frac{\partial \Psi}{\partial t} \right) \mathrm{d}\mathcal{V} \tag{2.12}$$

By inspection, for the integral to yield $\langle E \rangle = E = $ constant, one must have

$$-\frac{\hbar}{i} \frac{\partial \Psi}{\partial t} = E\Psi \tag{2.13}$$

Note that the direct substitution of Eq. (2.13) into Eq. (2.12) yields the desired result

$$\int_{\mathcal{V}} \Psi^* \left(-\frac{\hbar}{i} \frac{\partial \Psi}{\partial t} \right) \mathrm{d}\mathcal{V} = E \int_{\mathcal{V}} \Psi^* \Psi \, \mathrm{d}\mathcal{V} = E = \text{constant} \tag{2.14}$$

where use has been made of postulate 4. Equation (2.13) is in turn readily shown to have a general solution of the form

$$\Psi(x, y, z, t) = \psi(x, y, z)e^{-iEt/\hbar}$$ (2.15)

Next substituting Eq. (2.15) into the time-dependent Schrödinger equation [Eq. (2.9)], canceling the multiplicative factor $\exp(-iEt/\hbar)$ which appears in all terms, and slightly rearranging the resulting equation, one obtains

$$\nabla^2\psi + \frac{2m}{\hbar^2}[E - U(x, y, z)]\psi = 0$$ (2.16)

Equation (2.16) is referred to as the time-independent Schrödinger equation.

In essence, when the particle has a fixed total energy E, the time-dependence is completely specified and the problem reduces to solving the time-independent Schrödinger equation for the time-independent wavefunction $\psi = \psi(x, y, z)$. The remainder of the basic postulates can also be restated in terms of the time-independent wavefunction. Specifically,

(3′) ψ and $\nabla\psi$ must be finite, continuous, and single-valued for all values of x, y, and z.

(4′) Since $\Psi^*\Psi = \psi^*\psi$, $\psi^*\psi d\mathcal{V} = |\psi|^2 d\mathcal{V}$ is to be identified as the probability the particle will be found in the spatial volume element $d\mathcal{V}$. Likewise,

$$\int_{\mathcal{V}} \psi^*\psi \, d\mathcal{V} = 1$$ (2.17)

(5′) The expectation value of the system variable α is given by

$$\langle\alpha\rangle = \int_{\mathcal{V}} \psi^*\alpha_{op}\psi \, d\mathcal{V}$$ (2.18)

where α_{op} is the mathematical operator associated with α. α_{op} in Eq. (2.18) cannot explicitly depend on time.

All of the problems to be considered herein and most of the problems encountered in practice employ the foregoing time-independent formulation.

2.3 SIMPLE PROBLEM SOLUTIONS

The following simple problem solutions serve a threefold purpose: First, they help illustrate use of the quantum mechanical formalism and the interpretation of results. Second, the problems introduce additional formalism and a number of concepts that will prove to be of general utility. Finally, the problem solutions are of interest in themselves. The solutions, or reference to the results, can be found in the later chapters and in a number of solid-state device analyses.

The reader should be alerted to the fact that solutions to the first two problems violate quantum mechanical postulates (4') and (3'), respectively. The violations arise as a direct result of unrealistic idealizations that are introduced to make the problems tractable. Fortunately, neither violation poses a serious difficulty.

2.3.1 The Free Particle

PROBLEM SPECIFICATION:

The first problem to be addressed is the quantum mechanical characterization of a free particle. By definition, a free particle is an entity (say an electron) that finds itself alone in the universe. The particle is assumed to have a mass m and a fixed total energy E. Being alone, the particle will experience no forces and the potential energy of the system must likewise be a constant everywhere. The potential energy is of course arbitrary to within a constant, and we can therefore choose $U(x, y, z) = $ constant $= 0$. For simplicity, let us also take the universe to be one-dimensional.

SOLUTION:

To obtain the desired solution, we must clearly solve the time-independent Schrödinger equation. With $U(x, y, z) = 0$ and the particle restricted to one-dimensional motion ($\nabla^2 \rightarrow d^2/dx^2$), Eq. (2.16) simplifies to

$$\frac{d^2\psi}{dx^2} + \frac{2mE}{\hbar^2}\psi = 0 \tag{2.19}$$

By introducing the constant

$$k \equiv \sqrt{2mE/\hbar^2} \quad \left(\text{or equivalently, } E = \frac{\hbar^2 k^2}{2m}\right) \tag{2.20}$$

the equation to be solved can be manipulated into the form

$$\frac{d^2\psi}{dx^2} + k^2\psi = 0 \tag{2.21}$$

Equation (2.21) is a well-known differential equation whose general solution can be alternatively expressed in terms of sines and cosines, the hyperbolic functions, or exponentials. The last cited form of the solution is the most convenient in this particular problem. We therefore rapidly obtain the general solution

$$\psi(x) = A_+ e^{ikx} + A_- e^{-ikx} \tag{2.22}$$

where A_+ and A_- are solution constants. Furthermore, making use of Eq. (2.15), we conclude

$$\Psi(x, t) = A_+ e^{i(kx - Et/\hbar)} + A_- e^{-i(kx + Et/\hbar)} \qquad (2.23)$$

DISCUSSION:

The interpretation of the foregoing result is reasonably straightforward to those familiar with classical wave theory. In classical analyses dealing with electromagnetic waves, sound waves, and even mass waves on a vibrating string,

$e^{i(kx - \omega t)}$ ...corresponds to a wave traveling in the $+x$ direction

and

$e^{-i(kx + \omega t)}$ ...corresponds to a wave traveling in the $-x$ direction

where

$$\boxed{k \equiv \frac{2\pi}{\lambda}} \qquad \text{... is the } wavenumber \qquad (2.24)$$

and ω is the angular frequency of the traveling wave. Thus, by analogy, the free-particle wavefunction [Eq. (2.23)] is interpreted to be a traveling wave. If the particle is assumed to be moving in the $+x$ direction, it follows that $A_- = 0$. Likewise, A_+ would be zero for a free particle moving in the $-x$ direction. Also note that the introduction of $k = \sqrt{2mE/\hbar^2}$ in the wavefunction solution anticipated this constant being identified as the wavenumber.

Let us next see what can be deduced about the free particle itself. Assuming the particle is moving in the $+x$ direction, we note first of all that $\psi^*\psi dx$ $= A_+^* A_+ dx$ = constant for all values of x. Thus one has an equal probability of finding the particle in any dx spatial segment. The probability of finding the particle integrated over all space must be equal of course to unity according to postulate 4'. Integration of the probability density over all space is the usual means whereby one "normalizes" the wavefunction—i.e., determines the multiplicative constant in the wavefunction solution. However, a constant probability integrated between infinite limits technically requires $A_+^* A_+ = |A_+|^2$ and the associated wavefunction to become vanishingly small. The paradox here (which keeps us from determining A_+) arises because of the nonphysical size of the assumed universe. Simply limiting the size of the universe to some large but finite value would resolve the paradox without affecting any of the results or conclusions presented herein.

Another property of the free particle which we wish to investigate is its $+x$ direction momentum. Making use of postulate 5' and Table 2.1, we find

$$\langle p \rangle = \langle p_x \rangle = \int_{-\infty}^{\infty} \psi^* \frac{\hbar}{i} \frac{d\psi}{dx} \, dx = \hbar k \int_{-\infty}^{\infty} \psi^*\psi \, dx \qquad (2.25a)$$

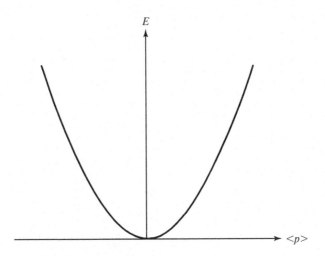

E

$\langle p\rangle$

Figure 2.3 Energy–momentum relationship for a free particle.

or, in light of postulate 4′ and Eq. (2.24),

$$[\langle p\rangle = \hbar k = h/\lambda] \tag{2.25b}$$

Note that Eq. (2.25b) is a restatement of the de Broglie relationship [Eq. (2.8)] and its derivation here was based solely on wave mechanical arguments. Thus the de Broglie free-particle hypothesis is implicitly contained in the postulates of wave mechanics.

Finally, let us examine the energy of the free particle. Expressing k in terms of the momentum using Eq. (2.25b), and substituting into the equivalent form of Eq. (2.20), one obtains

$$\left[E = \frac{\langle p\rangle^2}{2m}\right] \tag{2.26}$$

Now the energy of a *classical* free particle is equal to $mv^2/2$, $p = mv$, and therefore $E_{\text{classical}} = p^2/2m$. Hence, the quantum mechanical and classical free particles exhibit precisely the same energy–momentum relationship. This important $E–\langle p\rangle$ relationship is pictured in Fig. 2.3 for future reference. Also note that the quantum mechanical free particle, like its classical analog, can take on a continuum of energies: the energy of the free particle is not restricted to a quantized set of values.

2.3.2 Particle in a 1-D Box

PROBLEM SPECIFICATION:

The "particle-in-a-box" analysis, or characterization of a spatially confined entity, is more typical of wave mechanical problems. As pictured in Fig. 2.4(a), we envision a

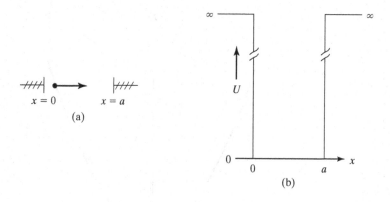

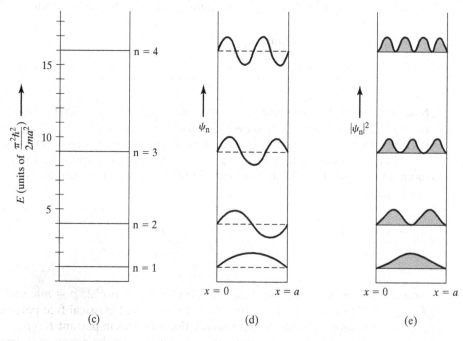

Figure 2.4   Particle in an infinitely deep one-dimensional potential well. (a) Spatial visualization of the particle confinement. (b) The assumed potential energy versus position dependence. (c) First four allowed energy levels. (d) Wavefunctions and (e) $|\psi|^2$ associated with the first four energy levels. $|\psi|^2$ is proportional to the probability of finding the particle at a given point in the potential well.

particle of mass m with fixed total energy E confined to a relatively small segment of one-dimensional space between $x = 0$ and $x = a$. In terms of the potential energy (see Fig. 2.4(b)), the particle may be viewed as being trapped in an infinitely deep one-dimensional potential well with $U(x) =$ constant for $0 < x < a$. Since the potential energy is arbitrary to within a constant, we can choose $U = 0$ for $0 < x < a$ without any loss in generality. Clearly, the formulation of the particle-in-a-box and free-particle problems are identical except for the size of the confining "box."

SOLUTION:

With $U = 0$ in the region of particle confinement, and given the one-dimensional nature of the problem, the time-independent Schrödinger equation again reduces to

$$\frac{d^2\psi}{dx^2} + k^2\psi = 0 \qquad \ldots 0 < x < a \tag{2.27}$$

where, as before,

$$k \equiv \sqrt{2mE/\hbar^2} \quad \text{or} \quad E = \frac{\hbar^2 k^2}{2m} \tag{2.28}$$

Since the particle cannot stray into the regions external to $0 < x < a$, the wavefunction in these regions must be identically zero. The wavefunction, however, must also be continuous at the region boundaries (postulate 3′), which imposes the boundary conditions

$$\psi(0) = 0 \tag{2.29a}$$

and

$$\psi(a) = 0 \tag{2.29b}$$

The general solution of Eq. (2.27), written in the form most convenient for this particular problem, is

$$\psi(x) = A \sin kx + B \cos kx \tag{2.30}$$

Next, applying the boundary conditions yields

$$\psi(0) = B = 0 \tag{2.31a}$$

and

$$\psi(a) = A \sin ka = 0 \tag{2.31b}$$

Other than the trivial $\psi = 0$ result obtained by setting $A = 0$, the Eq. (2.31b) condition is satisfied only when ka is a multiple of π. We therefore conclude that k is restricted to the values

$$[k = \mathbf{n}\pi/a \qquad \mathbf{n} = \pm 1, \ \pm 2, \ \pm 3, \ \cdots] \tag{2.32}$$

with the wavefunction corresponding to a given k (or $\mathbf{n}$) being

$$\left[\psi_{\mathbf{n}}(x) = A_{\mathbf{n}}\sin\frac{\mathbf{n}\pi x}{a} \right] \tag{2.33}$$

Likewise, making use of Eq. (2.28), we further conclude that the energy of the particle can only assume the quantized values

$$\left[E_{\mathbf{n}} = \frac{\mathbf{n}^2\pi^2\hbar^2}{2ma^2} \right] \tag{2.34}$$

DISCUSSION:

Perhaps the most striking feature of the foregoing results is the direct prediction of energy quantization, a quantization that arises in turn as a direct result of the particle confinement. The four lowest-lying energy levels are pictured in Fig. 2.4(c), while the wavefunctions and $|\psi|^2$ associated with the first four levels are shown, respectively, in Fig. 2.4(d) and (e).

The Fig. 2.4(d) wavefunction plots are highly suggestive of standing waves, and indeed the particle can be thought of as bouncing back and forth between the walls of the potential well. Since the particle periodically changes direction, it should come as no surprise that the expectation or average value of the particle's momentum, $\langle p \rangle = \langle p_x \rangle$, is precisely zero for all energy states. However, a standing wave can always be decomposed into two counterpropagating traveling waves. If this be done, the momentum associated with the component waves is readily shown to be $\mathbf{n}\pi\hbar/a$, where $\mathbf{n} > 0$ for $+x$ propagation and $\mathbf{n} < 0$ for $-x$ propagation. A plot of allowed particle energies [Eq. (2.34)] versus the counterpropagating wave momentum is shown in Fig. 2.5. Note that the discrete $E-p$ points derived from the particle-in-a-box analysis all lie along the continuous $E = \langle p \rangle^2/2m$ curve characteristic of a free particle. Conceptually increasing the width a of the potential well would cause the discrete points in Fig. 2.5 to move closer together and slide toward the origin of coordinates. In the limit where $a \to \infty$, the discrete points would form a quasi-continuum, thereby essentially replicating the free-particle curve. This is of course consistent with the particle-in-a-box becoming a free particle as $a \to \infty$.

We should point out that the $\mathbf{n}$-integer which appears in the k, $E_{\mathbf{n}}$, and $\psi_{\mathbf{n}}$ relationships is called a *quantum number*. Strictly speaking, as noted in Eq. (2.32), $\mathbf{n}$ can take on both positive and negative integer values. However, substituting into

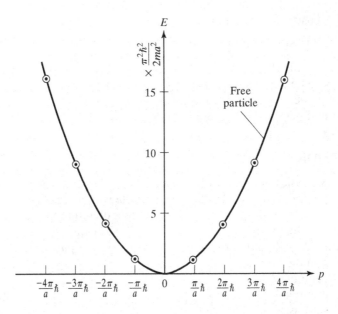

Figure 2.5 Allowed infinite-well particle energy versus counterpropagating wave momentum (discrete points) referenced against the free particle $E-\langle p\rangle$ relationship.

Eq. (2.33), one finds that the wavefunctions corresponding to the negative and positive values of the same whole number differ by only a minus sign—i.e., $\psi_{-n} = -\psi_{+n}$. As can be deduced from Eq. (2.11) or Eq. (2.18), when two wavefunctions differ only in sign, the expectation values for all system observables will be identical. In other words, one cannot physically distinguish between the $-\mathbf{n}$ and $+\mathbf{n}$ states; they actually refer to one and the same state. For this reason it is common practice to simply neglect the negative quantum numbers in the preceding analysis.

   In many quantum mechanics problems the analysis is considered complete when the allowed energy spectrum has been determined; the multiplicative constants in the wavefunction solutions are often left unspecified. Nevertheless, it is a relatively easy matter to normalize the wavefunction in the particular problem at hand. Substituting the $\psi_{\mathbf{n}}$ expression into Eq. (2.17), one rapidly deduces $A_{\mathbf{n}} = \sqrt{2/a}$. We should also mention that our wavefunction solutions for a particle in an infinitely deep potential well are not in strict compliance with postulate $3'$. As is evident from Fig. 2.4(d), the derivative of the wavefunctions is not continuous at the well boundaries ($d\psi/dx = 0$ for $x < 0$ and $x > a$). The source of the discrepancy is the unrealistic specification of an *infinite* well depth. Fortunately, it is possible to verify the results which we have presented by alternatively considering a particle in a finite potential well and examining the limit as the well depth goes to infinity. The finite potential well problem is addressed in the next subsection.

2.3.3   Finite Potential Well

PROBLEM SPECIFICATION:

The "particle in a finite potential well" is the last problem we will consider to illustrate the general procedures and concepts of quantum mechanics. The potential energy is taken to be as specified in Fig. 2.6, with $U(x) = 0$ for $0 < x < a$ and $U(x) = U_0$ for $x < 0$ and $x > a$. The particle under consideration has a mass m and a fixed total energy E.

SOLUTION $(0 < E < U_0)$:

A classical particle with an energy $0 < E < U_0$ would be confined to the envisioned potential well, while the same particle with an energy $E > U_0$ would be free to roam throughout all space. The behavior of a quantum mechanical particle is likewise distinctly different when $E < U_0$ and $E > U_0$. The quantitative analysis to be presented treats the $E < U_0$ situation, with comments on the particle's $E > U_0$ behavior being included in the ensuing discussion.

Beginning the $0 < E < U_0$ analysis, we note that the $x < 0, 0 < x < a$ and $x > a$ solutions of the time-independent Schrödinger equation must be handled on an individual basis. The subscripts $-$, 0, and $+$ will therefore be used to identify the wavefunctions and solution constants in these regions, respectively. Invoking the simplifications inherent in the problem, one obtains

$$\frac{d^2\psi_0}{dx^2} + k^2\psi_0 = 0 \qquad 0 < x < a \tag{2.35}$$

$$k \equiv \sqrt{2mE/\hbar^2} \tag{2.36}$$

and

$$\frac{d^2\psi_\pm}{dx^2} - \alpha^2\psi_\pm = 0 \qquad x < 0; x > a \tag{2.37}$$

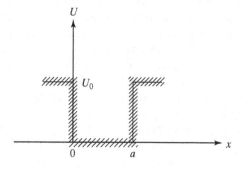

Figure 2.6   Finite potential well.

$$\alpha \equiv \sqrt{2m(U_0 - E)/\hbar^2} \qquad (0 < E < U_0) \qquad (2.38)$$

The general solutions to Eqs. (2.35) and (2.37) are

$$\psi_-(x) = A_-e^{\alpha x} + B_-e^{-\alpha x} \qquad \dots x < 0 \qquad (2.39a)$$

$$\psi_0(x) = A_0\sin kx + B_0\cos kx \qquad \dots 0 < x < a \qquad (2.39b)$$

$$\psi_+(x) = A_+e^{\alpha x} + B_+e^{-\alpha x} \qquad \dots x > a \qquad (2.39c)$$

Far from the well boundaries the wavefunction must vanish. Likewise, the wavefunction and its derivative must be continuous at $x = 0$ and $x = a$. These requirements translate into six boundary conditions—namely,

$$\psi_-(-\infty) = 0; \qquad \psi_+(+\infty) = 0 \qquad (\psi \to 0 \text{ as } x \to \pm\infty) \qquad (2.40a)$$

$$\psi_-(0) = \psi_0(0); \qquad \psi_0(a) = \psi_+(a) \qquad (\text{continuity of } \psi) \qquad (2.40b)$$

$$\left.\frac{d\psi_-}{dx}\right|_0 = \left.\frac{d\psi_0}{dx}\right|_0 ; \qquad \left.\frac{d\psi_0}{dx}\right|_a = \left.\frac{d\psi_+}{dx}\right|_a \qquad \left(\text{continuity of } \frac{d\psi}{dx}\right) \qquad (2.40c)$$

The Eq. (2.40a) boundary conditions can only be satisfied by setting $B_- = 0$ and $A_+ = 0$. The remaining boundary conditions give rise to a set of four simultaneous equations:

$$A_- = B_0 \qquad (2.41a)$$

$$A_0\sin ka + B_0\cos ka = B_+e^{-\alpha a} \qquad (2.41b)$$

$$\alpha A_- = kA_0 \qquad (2.41c)$$

$$kA_0\cos ka - kB_0\sin ka = -\alpha B_+e^{-\alpha a} \qquad (2.41d)$$

Seeking a solution to these equations, we note that B_0 can be readily expressed in terms of A_0 using Eqs. (2.41a) and (2.41c). After the B_0 expression is substituted into Eqs. (2.41b) and (2.41d), the resulting equations can be appropriately combined to obtain an equation involving only A_0. The net result is

$$A_0[(k^2 - \alpha^2)\sin ka - 2\alpha k\cos ka] = 0 \qquad (2.42)$$

To satisfy Eq. (2.42), either $A_0 = 0$ or the bracketed expression must be equal to zero. However, if $A_0 = 0$, all the other solution constants are likewise equal to zero, and one obtains the trivial $\psi = 0$ result. A non-trivial solution is therefore obtained if and only if

$$(k^2 - \alpha^2)\sin ka - 2\alpha k \cos ka = 0 \tag{2.43a}$$

or

$$\tan ka = \frac{2\alpha k}{k^2 - \alpha^2} \tag{2.43b}$$

To recast Eq. (2.43b) into a form more amenable to examination, let us introduce

$$\alpha_0 \equiv \sqrt{2mU_0/\hbar^2} \qquad (\alpha_0 = \text{constant}) \tag{2.44}$$

and

$$\xi \equiv E/U_0 \qquad (0 < \xi < 1) \tag{2.45}$$

One can then write

$$\alpha = \alpha_0 \sqrt{1 - \xi} \tag{2.46}$$
$$k = \alpha_0 \sqrt{\xi} \tag{2.47}$$

and therefore

$$\tan(\alpha_0 a \sqrt{\xi}) = \frac{2\sqrt{\xi(1 - \xi)}}{2\xi - 1} \tag{2.48}$$

Since α_0 and a are system constants, the normalized particle energy ξ is the only unknown in Eq. (2.48); the ξ values satisfying Eq. (2.48) correspond to the allowed particle energies. To solve Eq. (2.48) for the desired energy eigenvalues one must resort to numerical or graphical techniques. One approach would be to locate intersection points on superimposed plots of the tangent function versus ξ and $f(\xi) \equiv 2\sqrt{\xi(1 - \xi)}/(2\xi - 1)$ versus ξ. $f(\xi)$ versus ξ is plotted in Fig. 2.7.

DISCUSSION:

Let us first investigate the allowed particle energies as a function of potential well depth. For very shallow wells where $\alpha_0 a < \pi$ or $U_0 < \hbar^2\pi^2/2ma^2$, we find that there is one and only one allowed energy level. The $\tan\theta$ is a multi-branch function

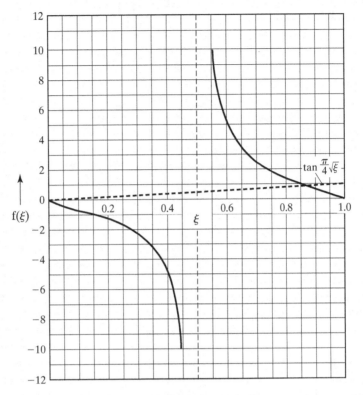

Figure 2.7   The $f(\xi)$ function used to determine the allowed energies of a particle in a finite potential well.

that monotonically increases from zero at $\theta = 0$ to $+\infty$ as $\theta \rightarrow \pi/2$, changes discontinuously to $-\infty$ at $\pi/2$, and then monotonically increases again to 0 at $\theta = \pi$. The described functional behavior is repeated for all subsequent $n\pi \leq \theta \leq (n + 1)\pi$ increments ($n = 1, 2, 3, \cdots$). If $\alpha_0 a < \pi$ the $\tan\theta$ is restricted to a portion of one repetitive unit and intercepts $f(\xi)$ at only one point, yielding the one allowed level. The specific case where $\alpha_0 a = \pi/4$ is illustrated in Fig. 2.7, from which one deduces the single allowed energy of $E = 0.87\,U_0$. This result is pictured in Fig. 2.8(a). Extending the preceding argument, one finds two allowed levels when $\pi \leq \alpha_0 a < 2\pi$, three allowed levels when $2\pi \leq \alpha_0 a < 3\pi$, four levels for $\alpha_0 a = 3\pi + \pi/4$ as shown in Fig. 2.8(b), etc. In the limit where $U_0 \rightarrow \infty$ (but E remains finite), the right-hand side of Eq. (2.43b) vanishes, $\tan ka = 0$, and one must have $ka = \mathbf{n}\pi$ ($\mathbf{n} = 1, 2, 3, \cdots$). Note that the limiting case solution here is identical to the result obtained in the infinite well analysis. Moreover, for a potential well of finite depth the energy levels always lie below the corresponding infinite well levels, with $E_n(\text{finite}) \rightarrow E_n(\text{infinite})$ at the lower energies (see Fig. 2.8(c)). Naturally, the deeper the finite well, the better the infinite well approximation for the lower-lying energy values.

Although we have not obtained explicit expressions for the wavefunction solution constants, it is still possible to deduce the shape of the wavefunctions from the general

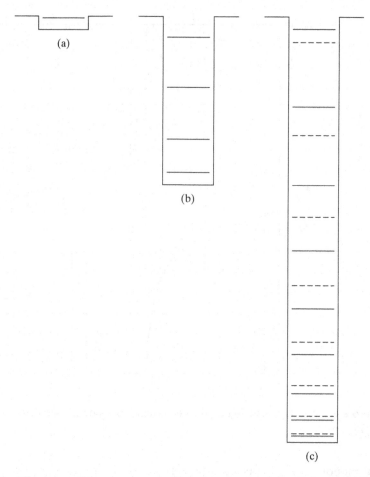

Figure 2.8   Allowed particle energies as a function of potential well depth. (a) Shallow well with single allowed level ($\alpha_0 a = \pi/4$). (b) Increase of allowed levels when $\alpha_0 a$ exceeds π ($\alpha_0 a = 3\pi + \pi/4$). (c) Comparison of the finite-well (———) and infinite-well (-----) energies ($\alpha_0 a = 8\pi + \pi/4$). All plots are drawn to scale.

form solutions and the earlier infinite well solutions. In particular, one would expect the wavefunction associated with the lowest energy state to exhibit the general form shown in Fig. 2.9(a). The wavefunction is roughly a half-period sinusoid within the well and falls exponentially to zero external to the well. The most interesting feature of the finite well wavefunction is its non-zero value external to the well. Since $|\psi|^2 \, dx$ is interpreted as the probability of finding the particle in a given dx region, a non-zero wave-function external to the well implies a finite probability of finding the particle outside the well in the classically "forbidden" region. (Classically, a particle with an energy $E < U_0$ cannot exist external to the well.) The significance of this observation is not readily apparent in relationship to the finite potential well problem. However, if the potential well is slightly

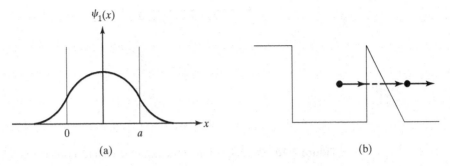

Figure 2.9   (a) Sketch of the wavefunction associated with the lowest energy state of a particle in a finite potential well. The sketch emphasizes the finite value of the wavefunction external to the well. (b) Visualization of tunneling through a thin barrier.

modified as envisioned in Fig. 2.9(b), the significance becomes self-evident. Given the finite value of the wavefunction in classically forbidden regions, the particle has a finite probability of "passing through" the Fig. 2.9(b) barrier and appearing as a free particle on the other side of the barrier. The quantum mechanical phenomenon of "passing through" a thin barrier, a phenomenon having no classical analog, is called *tunneling*. Tunneling provides the phenomenological basis for the tunnel diode and plays an important role in the operational behavior of a number of other solid-state devices.

As we have seen, a particle classically confined to a finite potential well $(0 < E < U_0)$ is subject to energy quantization. On the other hand, repeating the finite potential well analysis for particle energies $E > U_0$, energies which would permit a classical particle to roam throughout all space, one finds a continuum of allowed energies. These results are consistent with a pattern of results that can be formulated into a general rule. Whenever a particle is classically confined to a small spatial region, as was the case in the infinite potential well and finite $0 < E < U_0$ potential well problems, the particle will exhibit "bound" states with a discrete set of allowed energies. Conversely, whenever a particle is classically permitted to move unimpeded throughout a large spatial region, as exemplified by the free-particle and finite $E > U_0$ potential well problems, the particle will assume a continuum of allowed energies.

Finally, it should be mentioned that the finite potential well particle with $E > U_0$ does exhibit behavioral properties distinct from a free particle. Notably, since the wavefunction is different within and exterior to the potential well region, there is a finite probability that the particle will be reflected at the well boundaries as pictured in Fig. 2.10. Quantum mechanical reflection at a potential discontinuity comes into play, for example, in the detailed analysis of the Schottky diode current-voltage characteristics. Again, there is no classical analog for quantum mechanical reflection.

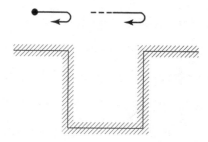

Figure 2.10   Visualization of quantum mechanical reflection.

REFERENCES

[1] Mathematical arguments justifying the form of Schrödinger's equation can be found in most texts on quantum mechanics and in a number of books treating solid-state device physics. For example, see J. L. Powell and B. Crasemann, *Quantum Mechanics*, Addison-Wesley Publishing Co., Inc., Reading, MA, 1961 (pp. 86-95) or H. E. Talley and D. G. Daugherty, *Physical Principles of Semiconductor Devices*, Iowa State University Press, Ames, 1976 (pp. 32–37).

[2] J. P. McKelvey, *Solid-State and Semiconductor Physics*, Harper and Row, New York, 1966; p. 75.

SOURCE LISTING

(1) H. Kroemer, *Quantum Mechanics for Engineering, Material Science, and Applied Physics*, Prentice Hall, Englewood Cliffs, NJ, 1994.

(2) J. P. McKelvey, *Solid State and Semiconductor Physics*, Harper and Row, New York, 1966.

(3) J. L. Powell and B. Crasemann, *Quantum Mechanics*, Addison-Wesley Publishing Co., Inc., Reading, MA, 1961.

(4) H. E. Talley and D. G. Daugherty, *Physical Principles of Semiconductor Devices*, Iowa State University Press, Ames, 1976.

PROBLEMS

2.1 The *exciton* is a hydrogen atom-like entity encountered in advanced semiconductor work. It consists of an electron bound to a $+q$ charged particle (a hole) of approximately equal mass. Bohr atom results can be used in computing the allowed energy states of the exciton provided the reduced mass, $m_r = m_+ m_- / (m_+ + m_-) \simeq m_0/2$, replaces the electron mass in the Bohr atom formulation. In addition, the distance between the components of the exciton is always such that there are intervening semiconductor atoms. Thus ε_0 in the Bohr formulation must also be replaced by $K_S \varepsilon_0$, where K_S is the semiconductor dielectric constant. Using $K_S = 11.8$, determine the ground state ($\mathbf{n} = 1$) energy of an exciton in Si.

2.2 Reflection High Energy Electron Diffraction (RHEED) has become a commonplace technique for probing the atomic surface structures of materials. Under vacuum conditions an electron beam is made to strike the surface of the sample under test at a glancing angle ($\theta \leq 10°$). The beam reflects off the surface of the material and subsequently strikes a phosphorescent screen. Because of the wave-like nature of the electrons, a diffraction pattern characteristic of the first few atomic layers is observed on the screen if the surface is flat and the material is crystalline. With a distance between atomic planes of $d = 5\,\text{Å}$, a glancing angle of 1°, and an operating de Broglie wavelength for the electrons of $2d\sin\theta$, compute the electron energy employed in the technique.

2.3 (a) Confirm, as pointed out in the text, that $\langle p_x \rangle = 0$ for all energy states of a particle in a 1-D box.

(b) Verify that the normalization factor for wavefunctions describing a particle in a 1-D box is $A_n = \sqrt{2/a}$.

(c) Determine $\langle x \rangle$ for all energy states of a particle in a 1-D box.

2.4 In examining the finite potential well solution, suppose we restrict our interest to energies where $\xi = E/U_0 \leq 0.01$ and permit a to become very large such that $\alpha_0 a \sqrt{\xi_{max}} \gg \pi$. Present an argument that concludes the energy states of interest will be very closely approximated by those of the infinitely deep potential well. (This approximation is invoked in Chapter 4 when we treat the density of states in a semiconductor crystal.)

2.5 The symmetry of a problem sometimes allows one to simplify the mathematics leading to a solution. If, for example, the $x = 0$ point in the finite potential well problem is moved to the middle of the well as pictured in Fig. P2.5, it becomes obvious that the wavefunction solution must be symmetric about $x = 0$; i.e., $\psi(-x) = \pm\psi(x)$.

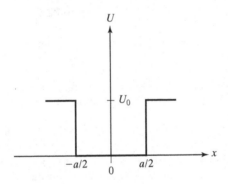

Figure P2.5

(a) Paralleling the development in the text, write down the equations to be solved in the various regions of the Fig. P2.5 potential well, the general solutions of those equations, and the boundary conditions to be applied.

(b) Simplify the general solutions by applying the boundary conditions at $x = \pm\infty$ and by invoking the symmetry requirement. You should now have *two* sets of equations—one set valid for even parity $[\psi(-x) = \psi(x)]$ and a second set valid for odd parity $[\psi(-x) = -\psi(x)]$.

(c) After applying the continuity boundary conditions, show that a non-trivial solution is obtained only if

$$\alpha = \begin{cases} k\tan(ka/2) & \ldots \text{for even parity} \\ -k\cot(ka/2) & \ldots \text{for odd parity} \end{cases}$$

(d) Confirm that the text Eq. (2.43b) and the part (c) expressions are equivalent.
HINT: $\tan ka = 2\tan(ka/2)/[1 - \tan^2(ka/2)]$.

2.6 A particle of mass m and fixed total energy E, where $0 < E < U_0$, is placed in the one-dimensional potential well pictured in Fig. P2.6.

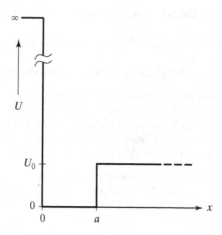

Figure P2.6

(a) Write down the simplified form of Schrodinger's equation appropriate for the various spatial regions.
(b) Indicate the general solutions to your part (a) equations.
(c) List the boundary conditions appropriate for the given problem.
(d) Establish the simultaneous equations that result by applying the part (c) boundary conditions.
(e) Obtain the equation that must be solved to determine the allowed particle energies.

2.7 The one-dimensional "triangular" potential well shown in Fig. P2.7 has been used in device work to model the near-surface region of semiconductors under certain biasing conditions. The well barrier is infinitely high at $x = 0$ and $U = ax$ for $x > 0$. We wish to investigate the solution for the allowed states of a particle of mass m and fixed total energy E placed into the triangular potential well.

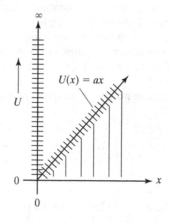

Figure P2.7

(a) Write down Schrödinger's equation for the specified problem. Let $k^2 = 2mE/\hbar^2$ and $x_e = E/a$.

(b) Assuming different wavefunction solutions (call these ψ_I and ψ_{II}) will be obtained for $0 < x < x_e$ and $x > x_e$, indicate the boundary conditions to be employed in solving the problem. (Actually, it is possible to obtain a single solution for ψ which automatically satisfies the boundary condition at $x = \infty$.)

(c) Sketch the expected general form of the ground state (lowest energy) wavefunction. Also indicate how you arrived at your sketch.

(d) Consulting an appropriate quantum mechanics text (for example, see S. Flügge, *Practical Quantum Mechanics I*, Springer-Verlag, Berlin, 1971, pp. 101–105), record the most convenient forms of the wavefunction solution for $0 < x \le x_e$ and $x \ge x_e$.

(e) Imposing the $x = 0$ boundary condition leads to the allowed energies

$$E_n \simeq (\hbar^2/2m)^{1/3}(3\pi a/2)^{2/3}(n + 3/4)^{2/3} \qquad \dots n = 0, 1, 2, \dots$$

Indicate the positioning of the five lowest energy levels in the triangular potential well on a dimensioned plot similar to Fig. 2.4c.

2.8 An excellent discussion of tunneling through a potential energy barrier is presented in Appendix B.2 of J. Singh, *Semiconductor Devices, Basic Principles*, John Wiley & Sons, Inc., New York, 2001.

(a) Briefly summarize the Singh discussion.

(b) Confirm the quoted expression for the transmission coefficient (T) through a square barrier by providing the missing mathematical steps.

2.9 The solution for the transmission coefficient (T) through a square barrier cited in Problem 2.8 and reproduced below is also valid for energies $E > U_0$, where U_0 is the barrier height.

$$T = \frac{4}{4\cosh^2 \alpha d + (\alpha/k - k/\alpha)^2 \sinh^2 \alpha d}$$

$$\text{where} \qquad k = \sqrt{2mE/\hbar^2}; \quad \alpha = \sqrt{2m(U_0 - E)/\hbar^2}$$

$$d = \text{barrier width}$$

(a) Introducing $\alpha = ik_0$ when $E > U_0$, revise the T-expression to eliminate $i = \sqrt{-1}$.

(b) Are there any finite values of k_0 and k where the transmission coefficient goes to unity? If so, cite the values. Physically, is there anything special about the cited T = 1 solution(s)?

(c) What is the limit of the T-expression as $E \to \infty$? What is the physical significance of this limiting case?

2.10 Consider a particle of mass m traveling from left to right over the potential well as pictured in text Fig. 2.10. Let E be the energy of the particle relative to the top of the well, $-U_0$ the depth of the well, and $-a < x < a$ the position of the well. Designate the three regions to the left, within, and to the right of the potential well as I, II, and III, respectively.

(a) Employing traveling-wave type solutions $[\psi = A\exp(ikx) + B\exp(-ikx)]$ in all three regions, establish a relationship for the transmittance (T) of the particle across the potential well, where $T \equiv |A_{III}/A_I|^2$. Note that because the particle will only be moving to the right in region III, $B_{III} = 0$.

(b) What is the limit of your T-expression as $a \to 0$? What is the physical significance of this limiting case?

(c) What is the limit of your T-expression as $E \to \infty$? What is the physical significance of this limiting case?

CHAPTER 3

Energy Band Theory

The results and concepts of band theory essential for performing device analyses are routinely presented in introductory texts. Extrapolating from the discrete energy states available to electrons in isolated atoms, it is typically argued that the interaction between atoms leads to the formation of energy bands, ranges of allowed electron energies, when the atoms are brought into close proximity in forming a crystal. The highest energy band containing electrons at temperatures above absolute zero is identified as the conduction band; the next-lower-lying band, separated from the conduction band by an energy gap on the order of an electron-volt in semiconductors, and mostly filled with electrons at temperatures of interest, is identified as the valence band. The carriers involved in charge transport or current flow are associated with filled states in the conduction band and empty states (holes) in the valence band, respectively.

With the quantum mechanical foundation established in the preceding chapter, we are able to present a straightforward development of the energy band model and a more sophisticated treatment of related concepts. Specifically, we will show that energy bands arise naturally when one considers the allowed energy states of an electron moving in a periodic potential—the type of potential present in crystalline lattices. The "essential" energy-band-related concepts found in introductory texts, such as the effective mass, will be expanded and examined in greater detail. Additional concepts encountered in advanced device analyses will also be presented and explained. The overall goal is to establish a working knowledge of the energy band description of electrons in crystals.

The chapter begins with a simplified formulation of the electron-in-a-crystal problem, and the introduction of a powerful mathematical theorem that is of use in dealing with periodic potentials. A one-dimensional analysis is then performed that leads to the direct prediction of energy bands. The one-dimensional result is used as a basis for introducing and discussing energy-band-related terms and concepts. The development is next generalized to three dimensions, with special emphasis being placed on the interpretation of commonly encountered informational plots and constructs.

3.1   PRELIMINARY CONSIDERATIONS

3.1.1   Simplifying Assumptions

Electrons moving inside a semiconductor crystal may be likened to particles in a three-dimensional box with a very complicated interior. In a real crystal at operational temperatures there will be lattice defects (missing atoms, impurity atoms, etc.), and the semiconductor atoms will be vibrating about their respective lattice points. To simplify the problem it will be our assumption that lattice defects and atom core vibrations lead to a second-order perturbation—i.e., we begin by considering the lattice structure to be perfect and the atoms to be fixed in position. Moreover, in our initial considerations we treat a one-dimensional analog of the actual crystal. This procedure yields the essential features of the electronic behavior while greatly simplifying the mathematics.

The potential energy function, $U(x)$, associated with the crystalline lattice is of course required before one can initiate the quantum mechanical analysis. The general form of the function can be established by considering the one-dimensional lattice shown in Fig. 3.1(a). Atomic cores (atomic nuclei plus the tightly bound core electrons) with a net charge $+Z'q$ and separated by a lattice constant a are envisioned to extend from $x = 0$ to $x = (N - 1)a$, where N is the total number of atoms in the crystal. If the atomic core-electron interaction is assumed to be purely coulombic, the attractive force between the $x = 0$ atomic core and an electron situated at an arbitrary point x would give rise to the potential energy versus x dependence pictured in Fig. 3.1(b). Adding the attractive force associated with the $x = a$ atomic core yields the potential energy dependence shown in Fig. 3.1(c). Ultimately, accounting for the electron interaction with all atomic cores, one obtains the periodic crystalline potential sketched in Fig. 3.1(d). This result, we should point out, neglects any non-core electron-electron interaction which may occur in the crystal. However, it is reasonable to assume that the non-core electron–electron interaction approximately averages out to zero, and that the allowed electron states within the crystal can be determined to first order by considering a single electron of constant energy E moving in a periodic potential well of the form pictured in Fig. 3.1(d).

3.1.2   The Bloch Theorem

The Bloch theorem is of great utility in quantum mechanical analyses involving periodic potentials. The theorem basically relates the value of the wavefunction within any "unit cell" of a periodic potential to an equivalent point in any other unit cell, thereby allowing one to concentrate on a single repetitive unit when seeking a solution to Schrödinger's equation. For a one-dimensional system the statement of the Bloch theorem is as follows:

IF $\qquad\qquad U(x)$ is periodic such that $U(x + a) = U(x)$

THEN $\qquad\qquad\qquad \psi(x + a) = e^{ika}\psi(x)$ $\qquad\qquad$ (3.1a)

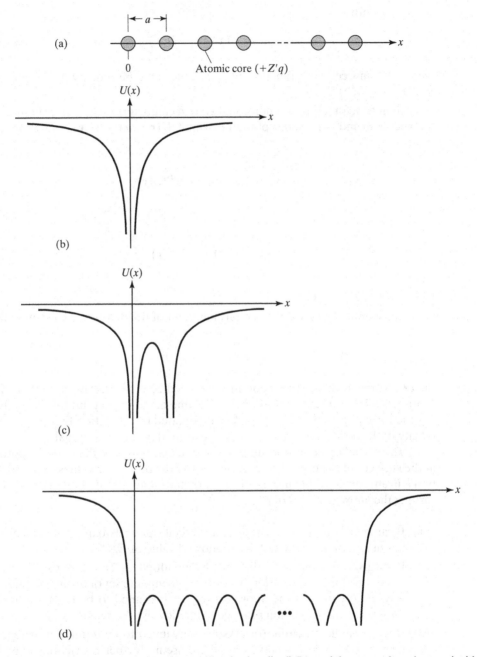

Figure 3.1 (a) One-dimensional crystalline lattice. (b–d) Potential energy of an electron inside the lattice considering (b) only the atomic core at $x = 0$, (c) the atomic cores at both $x = 0$ and $x = a$, and (d) the entire lattice chain.

or, equivalently,

$$\psi(x) = e^{ikx}u(x) \tag{3.1b}$$

where the unit cell wavefunction $u(x)$ has the same periodicity as the potential; i.e., $u(x + a) = u(x)$.

Similarly, for a three-dimensional system characterized by a translational symmetry vector $\boldsymbol{a}$ and a periodic potential where $U(\mathbf{r} + \boldsymbol{a}) = U(\mathbf{r})$, the Bloch theorem states

$$\psi(\mathbf{r} + \boldsymbol{a}) = e^{ik \cdot a}\psi(\mathbf{r}) \tag{3.2a}$$

or

$$\psi(\mathbf{r}) = e^{ik \cdot r}u(\mathbf{r}) \tag{3.2b}$$

where $u(\mathbf{r} + \boldsymbol{a}) = u(\mathbf{r})$.

Examining the one-dimensional statement of the theorem, please note that since

$$\psi(x + a) = e^{ik(x+a)}u(x + a) = e^{ika}e^{ikx}u(x) = e^{ika}\psi(x) \tag{3.3}$$

the alternative forms of the theorem are indeed equivalent. Also note that $\psi(x)$ itself is not periodic from unit cell to unit cell as one might expect intuitively. Rather, $\psi(x)$ has the form of a plane wave, $\exp(ikx)$, modulated by a function that reflects the periodicity of the crystalline lattice and the associated periodic potential.

The boundary conditions imposed at the end points of the periodic potential (or at the surfaces of the crystal) totally determine the permitted values of the Bloch function k in any given problem. Nevertheless, certain general statements can be made concerning the allowed values of k.

(1) It can be shown that, for a one-dimensional system, two and only two distinct values of k exist for each and every allowed value of E.

(2) For a given E, values of k differing by a multiple of $2\pi/a$ give rise to one and the same wavefunction solution. Therefore, a complete set of distinct k-values will always be obtained if the allowed k-values (assumed to be real) are limited to a $2\pi/a$ range. It is common practice to employ the Δk range $-\pi/a \leq k \leq \pi/a$.

(3) If the periodic potential (or crystal) is assumed to be infinite in extent, running from $x = -\infty$ to $x = +\infty$, then there are no further restrictions imposed on k other than k must be real—i.e., k can assume a continuum of values. k must be real if the crystal is taken to be infinite because the unit cell function $u(x)$ is well behaved for all values of x, while $\exp(ikx)$, and therefore $\psi(x)$, will blow up at either $-\infty$ or $+\infty$ if k contains an imaginary component.

(4) In dealing with crystals of finite extent, information about the boundary conditions to be imposed at the crystal surfaces may be lacking. To circumvent this

problem while still properly accounting for the finite extent of the crystal, it is commonplace to utilize what are referred to as *periodic boundary conditions*. The use of periodic boundary conditions is equivalent either to considering the ends of the crystal to be one and the same point, or to envisioning the lattice (Fig. 3.1(a)) to be in the form of a closed N-atom ring. For an N-atom ring with interatomic spacing a, one must have

$$\psi(x) = \psi(x + Na) = e^{ikNa}\psi(x) \tag{3.4}$$

which in turn requires

$$e^{ikNa} = 1 \tag{3.5}$$

or

$$k = \frac{2\pi\mathbf{n}}{Na} \qquad \ldots \mathbf{n} = 0, \pm1, \pm2, \cdots \pm N/2 \tag{3.6}$$

Thus, for a finite crystal, k can only assume a set of discrete values. Note that k has been limited to $-\pi/a \le k \le \pi/a$ in accordance with the discussion in (2) above, and the total number of distinct k-values is equal to N. Practically speaking, the large number of atoms N in a typical crystal will cause the Eq. (3.6) k-values to be very closely spaced, thereby yielding a quasi-continuum of allowed k-values.

3.2   APPROXIMATE ONE-DIMENSIONAL ANALYSIS

3.2.1   Kronig–Penney Model

Even with the simplifications presented in Section 3.1, the solution of Schrödinger's equation for an electron in a crystal remains quite formidable. Specifically, solution difficulties can be traced to the shape of the periodic potential. We therefore propose a further simplification, an idealization of the periodic potential as shown in Fig. 3.2. This idealization of the actual crystal potential is referred to as the Kronig–Penney model. Note that the modeled crystal is assumed to be infinite in extent.

The Kronig–Penney analysis must be considered a "classic"—required knowledge for anyone with a serious interest in devices. The value of the admittedly crude model stems from the fact that the associated analysis illustrates energy band concepts in a straightforward manner, with a minimum of math, and in a quasi-closed form. General features of the quantum mechanical solution can be applied directly to real crystals.

3.2.2   Mathematical Solution

The Kronig–Penney analysis closely parallels the finite potential well problem addressed in Subsection 2.3.3. We consider a particle, an electron, of mass m and fixed energy E subject to the periodic potential of Fig. 3.2(b). As in the finite potential well

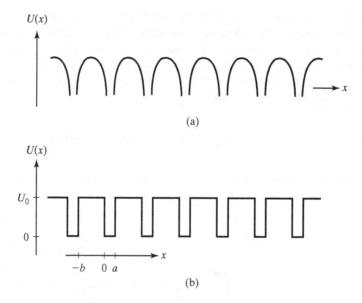

Figure 3.2   Kronig–Penney idealization of the potential energy associated with a one-dimensional crystalline lattice. (a) One-dimensional periodic potential. (b) Kronig–Penney model.

problem, we expect mathematically distinct solutions for the energy ranges $0 < E < U_0$ and $E > U_0$. Here, however, the two energy ranges will be handled simultaneously. The Bloch theorem of course relates the solution in an arbitrarily chosen unit cell of length $a + b$ to any other part of the crystal. For convenience let us choose $x = -b$ and $x = a$ as the unit cell boundaries, with the subscripts b and a identifying the wavefunctions and solution constants in the regions $-b < x < 0$ and $0 < x < a$, respectively. Schrödinger's equation, the equation to be solved in the two spatial regions, then assumes the form

$$\frac{d^2\psi_a}{dx^2} + \alpha^2\psi_a = 0 \qquad 0 < x < a \tag{3.7}$$

$$\alpha = \sqrt{2mE/\hbar^2} \tag{3.8}$$

and

$$\frac{d^2\psi_b}{dx^2} + \beta^2\psi_b = 0 \qquad -b < x < 0 \tag{3.9}$$

$$\beta = \begin{cases} i\beta_-; & \beta_- = \sqrt{2m(U_0 - E)/\hbar^2} & 0 < E < U_0 \tag{3.10a} \\[2ex] \beta_+; & \beta_+ = \sqrt{2m(E - U_0)/\hbar^2} & E > U_0 \tag{3.10b} \end{cases}$$

Written in the most convenient form, the general solutions to Eqs. (3.7) and (3.9) are

$$\psi_a(x) = A_a \sin\alpha x + B_a \cos\alpha x \qquad (3.11a)$$

$$\psi_b(x) = A_b \sin\beta x + B_b \cos\beta x \qquad (3.11b)$$

(We will subsequently replace the $\sin\beta x$ and $\cos\beta x$ with their hyperbolic equivalents when $\beta = i\beta_-$ is purely imaginary.) Now, the wavefunction and its derivative must be continuous at $x = 0$. Likewise, the wavefunction and its derivative evaluated at the cell boundaries must obey the periodicity requirements imposed by the Bloch theorem [Eq. (3.1a)]. These requirements translate into four boundary conditions:

$$\psi_a(0) = \psi_b(0) \qquad (3.12a)$$

$$\left.\frac{d\psi_a}{dx}\right|_0 = \left.\frac{d\psi_b}{dx}\right|_0 \qquad \left.\begin{array}{l} \text{Continuity} \\ \text{requirements} \end{array}\right. \qquad (3.12b)$$

$$\psi_a(a) = e^{ik(a+b)}\psi_b(-b) \qquad (3.12c)$$

$$\left.\frac{d\psi_a}{dx}\right|_a = e^{ik(a+b)}\left.\frac{d\psi_b}{dx}\right|_{-b} \qquad \left.\begin{array}{l} \text{Periodicity} \\ \text{requirements} \end{array}\right. \qquad (3.12d)$$

The Eq. (3.12) boundary conditions give rise to a set of four simultaneous equations:

$$B_a = B_b \qquad (3.13a)$$

$$\alpha A_a = \beta A_b \qquad (3.13b)$$

$$A_a \sin\alpha a + B_a \cos\alpha a = e^{ik(a+b)}[-A_b \sin\beta b + B_b \cos\beta b] \qquad (3.13c)$$

$$\alpha A_a \cos\alpha a - \alpha B_a \sin\alpha a = e^{ik(a+b)}[\beta A_b \cos\beta b + \beta B_b \sin\beta b] \qquad (3.13d)$$

Equations. (3.13a) and (3.13b) can be used to readily eliminate A_b and B_b in Eqs. (3.13c) and (3.13d), yielding

$$A_a[\sin\alpha a + (\alpha/\beta)e^{ik(a+b)}\sin\beta b] + B_a[\cos\alpha a - e^{ik(a+b)}\cos\beta b] = 0 \qquad (3.14a)$$

$$A_a[\alpha\cos\alpha a - \alpha e^{ik(a+b)}\cos\beta b] + B_a[-\alpha\sin\alpha a - \beta e^{ik(a+b)}\sin\beta b] = 0 \qquad (3.14b)$$

Paralleling the finite potential well problem, we could next proceed to eliminate B_a between the two remaining equations. It is expedient in the present situation, however, to make use of a well-known mathematical result—namely, that a set of n homogeneous equations linear in n unknowns has a non-trivial solution (a solution where the unknowns are non-zero) only when the determinant formed from the coefficients of the unknowns is equal to zero. Thus the first bracketed expression in Eq. (3.14a) times the second bracketed expression in Eq. (3.14b) minus the first bracketed expression in Eq. (3.14b) times the second bracketed expression in Eq. (3.14a) must be equal to zero.[†] Performing the required cross-multiplication and simplifying the result as much as possible, one obtains

$$-\frac{\alpha^2 + \beta^2}{2\alpha\beta}\sin\alpha a \,\sin\beta b + \cos\alpha a \,\cos\beta b = \cos k(a + b) \qquad (3.15)$$

Finally, reintroducing $\beta = i\beta_-$ for $0 < E < U_0$ and $\beta = \beta_+$ for $E > U_0$, noting $\sin(ix) = i\sinh x$ and $\cos(ix) = \cosh x$, and defining

$$\alpha_0 \equiv \sqrt{2mU_0/\hbar^2} \qquad (3.16)$$

$$\xi \equiv E/U_0 \qquad (3.17)$$

such that $\alpha = \alpha_0\sqrt{\xi}$, $\beta_- = \alpha_0\sqrt{1 - \xi}$ and $\beta_+ = \alpha_0\sqrt{\xi - 1}$, we arrive at the result

$$\frac{1 - 2\xi}{2\sqrt{\xi(1 - \xi)}}\sin\alpha_0 a\sqrt{\xi}\,\sinh\alpha_0 b\sqrt{1 - \xi} + \cos\alpha_0 a\sqrt{\xi}\,\cosh\alpha_0 b\sqrt{1 - \xi}$$
$$= \cos k(a + b) \qquad \ldots 0 < E < U_0 \qquad (3.18a)$$

$$\frac{1 - 2\xi}{2\sqrt{\xi(\xi - 1)}}\sin\alpha_0 a\sqrt{\xi}\,\sin\alpha_0 b\sqrt{\xi - 1} + \cos\alpha_0 a\sqrt{\xi}\,\cos\alpha_0 b\sqrt{\xi - 1}$$
$$= \cos k(a + b) \qquad \ldots E > U_0 \qquad (3.18b)$$

Other than system constants, the left-hand sides of Eqs. (3.18a) and (3.18b) depend only on the energy E, while the right-hand sides depend only on k. Consequently, Eqs. (3.18) specify the allowed values of E corresponding to a given k.

3.2.3   Energy Bands and Brillouin Zones

We are at long last in a position to confirm that energy bands arise naturally when one considers the allowed energy states of an electron moving in a periodic potential.

[†]If so inclined, we could have applied the determinant rule to the original set of four equations and to the four simultaneous equations encountered in the finite potential well problem. The end result is the same.

Because the crystal under analysis was assumed to be infinite in extent, the k in Eqs. (3.18) can assume a continuum of values and must be real. (This follows from observation #3 presented in Subsection 3.1.2.) The $\cos k(a + b)$ can therefore take on any value between -1 and $+1$. E-values which cause the left-hand side of Eq. (3.18a) or (3.18b), call this $f(\xi)$, to lie in the range $-1 \leq f(\xi) \leq 1$ are then the allowed system energies.

For a given set of system constants, the allowed values of E can be determined by graphical or numerical methods. To illustrate the graphical procedure and the general nature of the results, we have plotted $f(\xi)$ versus ξ in Fig. 3.3 for the specific case where $\alpha_0 a = \alpha_0 b = \pi$. From Fig. 3.3, $f(\xi)$ is seen to be an oscillatory-type function that alternately drops below -1 and rises above $+1$. This same behavior is observed for any set of system constants. Thus, as we have anticipated, there are extended ranges of allowed system energies (the shaded regions in Fig. 3.3). The ranges of allowed energies are called *energy bands;* the excluded energy ranges, *forbidden gaps* or *band gaps.* Relative to the crystal potential, the energy bands formed inside a crystal with $\alpha_0 a = \alpha_0 b = \pi$ would be roughly as envisioned in Fig. 3.4.

If the allowed values of energy are plotted as a function of k, one obtains the E-k diagram shown in Fig. 3.5. In constructing this plot the allowed values of k were limited to the $2\pi/(a + b)$ range between $-\pi/(a + b)$ and $+\pi/(a + b)$. As noted previously in Subsection 3.1.2, confining the allowed k-values to a $2\pi/$(cell length) range properly

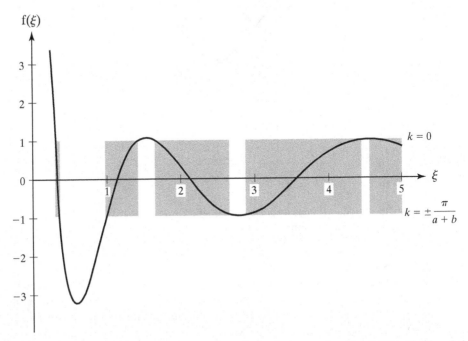

Figure 3.3 Graphical determination of allowed electron energies. The left-hand side of the Eqs. (3.18) Kronig–Penney model solution is plotted as a function of $\xi = E/U_0$. The shaded regions where $-1 \leq f(\xi) \leq 1$ identify the allowed energy states $(\alpha_0 a = \alpha_0 b = \pi)$.

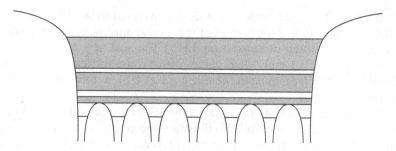

Figure 3.4   Visualization of the energy bands in a crystal.

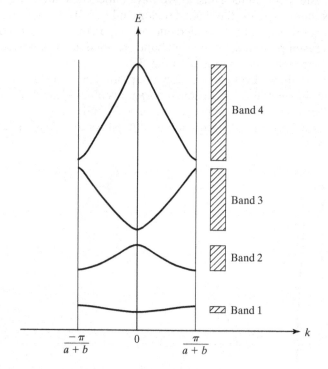

Figure 3.5   Reduced-zone representation of allowed E-k states in a one-dimensional crystal (Kronig–Penney model with $\alpha_0 a = \alpha_0 b = \pi$).

accounts for all distinct k-values—a complete set of distinct solutions lies within the cited range. Increasing or decreasing k in Eqs. (3.18) by a multiple of $2\pi/(a + b)$ has no effect on the allowed system energy and thus an E-k solution lying outside the chosen range simply duplicates one of the E-k solutions inside the chosen range. Also note in Fig. 3.5 that, consistent with observation #1 presented in the Bloch theorem discussion, there are two and only two k-values associated with each allowed energy.

    E-k diagrams are very important in the characterization of materials and we will have a great deal to say about their interpretation and use. At this point, however, we would merely like to point out that the energy band slope, dE/dk, is zero at the k-zone boundaries; i.e., $dE/dk = 0$ at $k = 0$ and $k = \pm\pi/(a + b)$ in Fig. 3.5. This is a feature common to all E-k plots, even those characterizing real materials.

Whereas Fig. 3.5 exemplifies the preferred and most compact way of presenting actual E-k information, valuable insight into the electron-in-a-crystal solution can be gained if the E-k results are examined from a somewhat different viewpoint. Specifically, instead of restricting k to the values between $\pm\pi/(a + b)$, one could alternatively associate increasing values of allowed E deduced from Eqs. (3.18) with monotonically increasing values of $|k|$. This procedure yields the E-k diagram shown in Fig. 3.6. As indicated in Fig. 3.6, the same result could have been achieved by starting with Fig. 3.5 and translating half-segments of the various bands along the k-axis by a multiple of $2\pi/(a + b)$. k-value solutions differing by $2\pi/(a + b)$ are of course physically indistinct, and therefore Figs. 3.5 and 3.6 are totally equivalent. When presented in the Fig. 3.6 format, however, the relationship between the periodic potential and the free-particle E-k solutions becomes obvious. The periodic potential introduces a perturbation

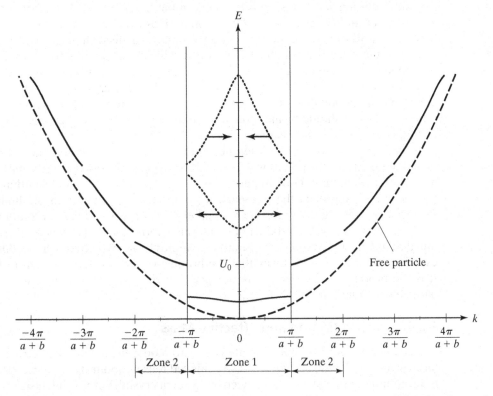

Figure 3.6 Extended-zone representation of allowed E-k states in a one-dimensional crystal (Kronig–Penney model with $\alpha_0 a = \alpha_0 b = \pi$). Shown for comparison purposes are the free-particle E-k solution (dashed line) and selected bands from the reduced-zone representation (dotted lines). Arrows on the reduced-zone band segments indicate the directions in which these band segments are to be translated to achieve coincidence with the extended-zone representation. Brillouin zones 1 and 2 are also labeled on the diagram.

that segments and distorts the free-particle solution. The modification is greatest at the lower energies, with the two solutions essentially merging at the higher energies. This seems reasonable from an intuitive standpoint, since the greater the electron energy, the smaller the relative importance of the periodic potential within the crystal.

Figure 3.6 also allows us to introduce relevant band-theory terminology. Those k-values associated with a given energy band are said to form a *Brillouin zone*. Brillouin zones are numbered consecutively beginning with the lowest energy band. The first Brillouin zone in Fig. 3.6 runs from $-\pi/(a + b)$ to $\pi/(a + b)$, the second from $\pm\pi/(a + b)$ to $\pm2\pi/(a + b)$, etc. For obvious reasons, an E-k diagram of the type presented in Fig. 3.6 is called an *extended-zone representation*. When the bands are all folded back into the first Brillouin zone as in Fig. 3.5, the diagram is called a *reduced-zone representation*.

To conclude this subsection, a comment is in order concerning the interpretation of the Bloch parameter k, the k found in E-k diagrams. For a free particle, k is of course the wavenumber and $\hbar k = \langle p \rangle$ is the particle momentum. Given the similarity between the free-particle solution and the extended-zone representation of the periodic potential solution, it is not surprising that k in the latter case is also referred to as the wavenumber and $\hbar k$ as the *crystal momentum*. However, as the addition of the word "crystal" implies, $\hbar k$ is not the actual momentum of an electron in a crystal, but rather a momentum-related constant of the motion which incorporates the crystal interaction. One might have suspected the $\hbar k$ in crystal plots to be different from the actual momentum since $\pm2\pi/(\text{cell length})$ can be added to the crystal momentum without modifying the periodic potential solution. $\hbar$ times the k appearing in the reduced-zone representation, it should be mentioned, is often called the reduced crystal momentum, or simply the *reduced momentum*.

Although accepting the difference between $\hbar k$(crystal) and the actual momentum, one might still wonder how $\pm2\pi/(\text{cell length})$ can be added to k(crystal) without modifying the solution. How, in particular, can the periodic potential solution with an "adjustable" k approach the free-particle solution with a fixed k in the limit where $E \gg U_0$? In this regard it must be remembered that the wavefunction solution for an electron in a crystal is the product of two terms, $\exp(ikx)$ and $u(x)$, where $u(x)$ is also a function of k. Increasing or decreasing k by a multiple of $2\pi/(\text{cell length})$ modifies both $\exp(ikx)$ and $u(x)$ in such a way that the product of the two terms is left unchanged. It is the product of the two terms, not just $\exp(ik_{\text{crystal}}x)$, that approaches the free-particle solution in the $E \gg U_0$ limit.

3.2.4   Particle Motion and Effective Mass

It was noted earlier that our ultimate goal is to model the *action* of electrons in crystals. The energy band solution we have achieved tells us about the allowed energy and reduced momentum states of an electron inside a crystal, but it is intrinsically devoid of action information. By assuming that the electron has a given energy E, we are automatically precluded from determining anything about the time evolution of the particle's position. Likewise, having specified k with absolute precision, the best we can do is compute the probability of finding the particle inside the various regions of the crystal.

The cited inability to deduce information about the position and motion of the particle when the energy and momentum are precisely specified is fundamental to the formulation of quantum mechanics. The fundamental property to which we refer is usually stated in terms of the *Heisenberg uncertainty principle*. The uncertainty principle observes that there is a limitation to the precision with which one can simultaneously determine conjugate dynamical variables. Specifically, for the E-t and p_x-x variable pairs, the precision is limited to

$$\Delta E \Delta t \geq \hbar \qquad (3.19a)$$

$$\Delta p_x \Delta x \geq \hbar \qquad (3.19b)$$

where the Δ in Eqs. (3.19) is to be read "the uncertainty in." Clearly, if the E of a particle is specified with absolute precision, the uncertainty in t is infinite; one is precluded from determining anything about the time evolution of the particle's position. Consequently, a *superposition* of fixed-E wavefunction solutions must be used to describe a particle if it is experimentally or conceptually confined to a given segment of a crystal at a given instant in time. In other words, to address the question of particle motion inside the crystal one must work with "wavepackets."

The wavepacket is the quantum mechanical analog of a classical particle localized to a given region of space. The wavepacket, literally a packet of waves, consists of a linear combination of constant-E wavefunction solutions closely grouped about a peak or center energy. The wavefunctions are assumed to be combined in such a way that the probability of finding the represented particle in a given region of space is unity at some specified time. Completely analogous to the Fourier series expansion of an electrical voltage pulse, the smaller the width of the wavepacket, the more constant-E solutions (analogous to Fourier components) of appreciable magnitude needed to accurately represent the wavepacket.

Reaction of the wavepacket to external forces and its spatial evolution with time provide the sought-after equation of motion for an electron in a crystal. Corresponding to the center of mass of a classical particle moving with a velocity v, one can speak of the wavepacket's center moving with a group velocity $v_g = dx/dt$. For a packet of traveling waves with center frequency ω and center wavenumber k, classical wave theory yields the dispersion relationship

$$v_g = \frac{d\omega}{dk} \qquad (3.20)$$

As is most readily evident from a comparison of free-particle and traveling-wave expressions given in Subsection 2.3.1, $E/\hbar$ in the quantum mechanical formulation replaces ω in the classical formulation. The wavepacket group velocity is therefore concluded to be

$$v_g = \frac{1}{\hbar} \frac{dE}{dk} \qquad (3.21)$$

where E and k are interpreted as the center values of energy and crystal momentum, respectively.

We are now in a position to consider what happens when an "external" force F acts on the wavepacket. F could be any force other than the crystalline force associated with the periodic potential. The crystalline force is already accounted for in the wavefunction solution. The envisioned force might arise, for example, from dopant ions within the crystal or could be due to an externally impressed electric field. The force F acting over a short distance dx will do work on the wavepacket, thereby causing the wavepacket energy to increase by

$$dE = F dx = F v_g dt \tag{3.22}$$

We can therefore assert

$$F = \frac{1}{v_g}\frac{dE}{dt} = \frac{1}{v_g}\frac{dE}{dk}\frac{dk}{dt} \tag{3.23}$$

or, making use of the group-velocity relationship,

$$F = \frac{d(\hbar k)}{dt} \tag{3.24}$$

Next, differentiating the group-velocity relationship with respect to time, we find

$$\frac{dv_g}{dt} = \frac{1}{\hbar}\frac{d}{dt}\left(\frac{dE}{dk}\right) = \frac{1}{\hbar^2}\frac{d^2E}{dk^2}\frac{d(\hbar k)}{dt} \tag{3.25}$$

which when solved for $d(\hbar k)/dt$ and substituted into Eq. (3.24) yields

$$\boxed{F = m^*\frac{dv_g}{dt}} \tag{3.26}$$

$$\boxed{m^* \equiv \frac{1}{\dfrac{1}{\hbar^2}\dfrac{d^2E}{dk^2}}} \tag{3.27}$$

The foregoing is a very significant result; its importance cannot be overemphasized. Equation (3.26) is identical to Newton's second law of motion except that the actual particle mass is replaced by an *effective mass* m^*. This implies that the motion of electrons in a crystal can be visualized and described in a quasi-classical manner. In most instances the electron can be thought of as a "billiard ball," and the electronic motion can be modeled using Newtonian mechanics, provided that one accounts for the effect of crystalline forces and quantum mechanical properties through the use of the effective-mass factor. Practically speaking, because of the cited simplification, device analyses can often be completed with minimal direct use of the quantum mechanical formalism. The effective-mass relationship itself, Eq. (3.27), also underscores the practical importance of the *E-k* diagrams discussed previously. Having established the crystal band structure or *E-k* relationship, one can determine the effective mass exhibited by the carriers in a given material.

Mathematically, the effective mass is inversely proportional to the curvature of an *E* versus *k* plot. It is therefore possible to deduce certain general facts about the effective mass from an *E-k* diagram simply by inspection. Consider, for example, the two band segments pictured in Fig. 3.7. In the vicinity of the respective energy minima, the curvature of segment (b) is greater than the curvature of segment (a). With $(d^2E/dk^2)_b > (d^2E/dk^2)_a$, one concludes that $m_a^* > m_b^*$. This example illustrates that the relative size of the carrier m^*'s in different bands can be readily deduced by inspection.

Consider next the band segment of the Kronig–Penney type reproduced in Fig. 3.8(a). Using graphical techniques, one finds the first and second derivatives of the

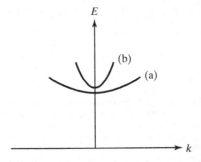

Figure 3.7 Hypothetical band segments used to illustrate how the relative magnitudes of the effective masses can be deduced from curvature arguments. By inspection, $m_a^* > m_b^*$ near $k = 0$.

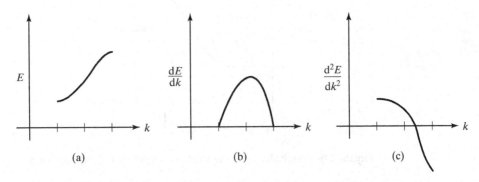

Figure 3.8 Deducing the sign of the effective-mass factor. (a) Sample band segment. (b) Graphically deduced first derivative and (c) second derivative of energy with respect to wavenumber.

band segment are roughly as sketched in Fig. 3.8(b and c), respectively. From Fig. 3.8(c) and Eq. (3.27), one in turn concludes that $m^* > 0$ near the band-energy minimum and $m^* < 0$ near the band-energy maximum. Since the shape of the Fig. 3.8(a) band segment is fairly typical of the band contours encountered in real materials, the preceding result is quite universal:

> m^* is positive near the bottoms of all bands.
>
> m^* is negative near the tops of all bands.

A negative effective mass simply means that, in response to an applied force, the electron will accelerate in a direction opposite to that expected from purely classical considerations.

In general, the effective mass of an electron is a function of the electron energy E. This is clearly evident from Fig. 3.8(c). However, near the top or bottom band edge—the region of the band normally populated by carriers in a semiconductor[†] —the E-k relationship is typically parabolic; i.e., as visualized in Fig. 3.9,

$$E - E_{\text{edge}} \simeq (\text{constant})\,(k - k_{\text{edge}})^2 \qquad (3.28)$$

and therefore

$$\frac{\mathrm{d}^2 E}{\mathrm{d}k^2} \simeq \text{constant} \qquad \dots E \text{ near } E_{\text{edge}} \qquad (3.29)$$

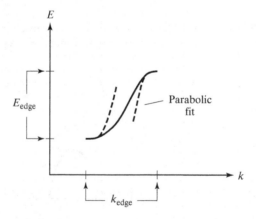

Figure 3.9 Parabolic nature of the bands near energy extrema points.

[†] For confirmation of this assertion, see Subsection 4.3.1

Thus, *carriers in a crystal with energies near the top or bottom of an energy band typically exhibit a CONSTANT (energy-independent) effective mass.* The usefulness of the effective-mass approximation would be questionable, and the correlation with classical behavior decidedly more complicated, if the effective carrier masses were not constant near the band edges.

3.2.5   Carriers and Current

The reader is no doubt aware that two types of carriers are present in semiconductors; namely, conduction-band electrons and valence-band holes. In this subsection we approach the origin and identification of these carriers from a band-theory standpoint. The band-theory approach provides additional insight, allowing one to answer certain questions that are difficult to address on an elementary level.

As a basis for discussion we consider a large one-dimensional crystal maintained at room temperature. The band structure of the crystal is assumed to be generally characterized by the Kronig–Penney model solution of Fig. 3.5. Because the crystal in the Kronig–Penney analysis was taken to be infinite in extent, there are an infinite number of k-states associated with each band in the cited solution. Restricting the number of atoms to some large but finite number, N, would have little effect on the energy-band structure. The number of distinct k-values in each band, however, would then be limited to N and spaced at $2\pi/N(a + b)$ intervals in accordance with observation #4 presented in Subsection 3.1.2. We assume this to be the case. For the sake of discussion, we further assume that each atom contributes two electrons to the crystal as a whole (there are two non-core electrons per atom), giving a grand total of $2N$ electrons to be distributed among the allowed energy states. At temperatures approaching zero Kelvin, the electrons would assume the lowest possible energy configuration: the available $2N$ electrons would totally fill the two lowest energy bands which contain N allowed states each. At room temperature, however, a sufficient amount of thermal energy is available to excite a limited number of electrons from the top of the second band into the bottom of the third band. We therefore conclude the electronic configuration within our crystal will be roughly as pictured in Fig. 3.10.

If a voltage is impressed across the crystal, a current will flow through the crystal and into the external circuit. Let us examine the contributions to the observed current from the various bands. The fourth band is of course totally devoid of electrons. Without "carriers" to transport charge there can be no current: *totally empty bands do not contribute to the charge-transport process.* Going to the opposite extreme, consider next the first band, where all available states are occupied by electrons. The individual electrons in this band can be viewed as moving about with velocities $v(E) = (1/\hbar)(dE/dk)$. However, because of the band symmetry and the filling of all available states, for every electron with a given $|v|$ traveling in the $+x$ direction, there will be another electron with precisely the same $|v|$ traveling in the $-x$ direction. Please note that this situation cannot be changed by the applied force: the absence of empty states precludes a modification of the electron velocity distribution within the band. Consequently, the first band, like the fourth band, does not contribute to the observed current: *totally filled bands do not contribute to the charge-transport process.*

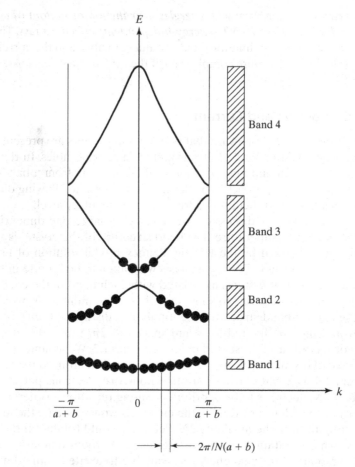

Figure 3.10   Filled and empty electronic states in the envisioned one-dimensional N-atom crystal maintained at room temperature. Each atom is assumed to contribute two electrons to the electronic configuration, and equilibrium conditions prevail. (Electrons are represented by filled circles.)

It follows from the foregoing discussion that only partially filled bands can give rise to a net transport of charge within the crystal. Under equilibrium conditions (Fig. 3.10), the filled-state distribution in the partially filled bands is symmetric about the band center and no current flows. Under the influence of an applied field, however, the filled-state distribution becomes skewed as envisioned in Fig. 3.11(a), and a current contribution is to be expected from both the second and third bands.

For the nearly empty third band, the contribution to the overall current will be

$$I_3 = -\frac{q}{L} \sum_{i(\text{filled})} v_i \qquad (3.30)$$

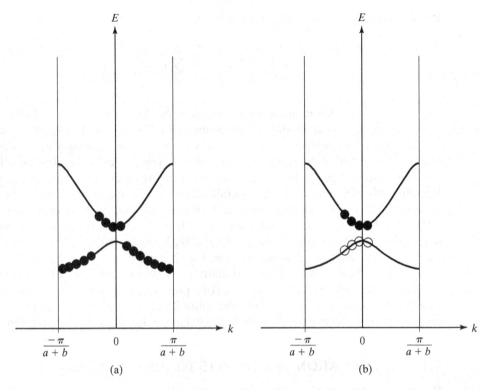

Figure 3.11   (a) The skewed filled-state distribution under steady-state conditions subsequent to the application of an external force. (b) Introduction of the hole. Alternative description of the electronic configuration in the lower energy band.

where I_3 is the current, L is the length of the one-dimensional crystal, and the summation is understood to be over all filled states. The Eq. (3.30) result is clearly analogous to the current attributed to conduction-band electrons in real crystals.

For the nearly filled second band, one could likewise write

$$I_2 = -\frac{q}{L} \sum_{i(\text{filled})} v_i \tag{3.31}$$

The summation in this case is rather cumbersome since it extends over a very large number of states. To simplify the result, we note that (as established previously) the summation of v_i over all states in a band is identically zero:

$$\sum_{i(\text{all})} v_i = 0 \tag{3.32}$$

It is therefore permissible to write

$$I_2 = -\frac{q}{L} \sum_{i(\text{filled})} v_i + \frac{q}{L} \sum_{i(\text{all})} v_i = \frac{q}{L} \sum_{i(\text{empty})} v_i \qquad (3.33)$$

where v_i in the last summation is understood to be the velocity $v(E) = (1/\hbar)\,(dE/dk)$ associated with the empty states. Interestingly, the form of the last result is what one would expect if a *positively* charged entity were placed in the empty electronic states and the remainder of the states in the band were considered to be *unoccupied* by the positively charged entity. The suggested conceptual revision is pictured in Fig. 3.11(b). Pursuing this idea, one finds the overall motion of the electrons in the nearly filled band can likewise be described by considering just the empty electronic states—PROVIDED THAT the effective mass associated with the empty states is taken to be the negative of the m^* deduced from Eq. (3.27). We know from the effective-mass discussion, however, that the m^* deduced from Eq. (3.27) is itself negative near the tops of energy bands. Thus, conceptually and mathematically, we can model the action of the electrons in a nearly filled band in terms of a positively charged entity with *positive* effective mass occupying empty electronic states. The cited entity is called a *hole* and the arguments we have presented apply to the valence-band holes in real materials.

3.3   EXTRAPOLATION OF CONCEPTS TO THREE DIMENSIONS

Basic energy-band concepts were introduced in the previous section using a simplified one-dimensional model of a crystalline lattice. In this section we examine the modifications required in extending these basic concepts to three dimensions and real crystals. No attempt will be made to present a detailed derivation of the 3-D results. Rather, emphasis will be placed on identifying the special features introduced by the 3-D nature of real materials and on interpreting oft-encountered plots containing band-structure information. The specific topics to be addressed are Brillouin zones, *E-k* diagrams, constant-energy surfaces, effective mass, and the band gap energy.

3.3.1   Brillouin Zones

The band structure of a one-dimensional lattice was described in terms of a one-dimensional or scalar k. 1-D Brillouin zones were in turn simply lengths or ranges of k associated with a given energy band. In progressing to the band-structure description of three-dimensional space lattices, the Bloch wavenumber becomes a vector and Brillouin zones become volumes. Specifically, a Brillouin zone (3-D) is the volume in k-space enclosing the set of k-values associated with a given energy band.

The first Brillouin zone for materials crystallizing in the diamond and zincblende lattices (Si, GaAs, etc.) is shown in Fig. 3.12. Geometrically, the zone is an octahedron which has been truncated by $\{100\}$ planes $2\pi/a$ from the zone center, a being the cubic lattice constant. The markings in the figure are group-theory symbols for high-symmetry points. (Don't panic. Familiarity with group theory is not required.) The most widely employed of the group-theory symbols are

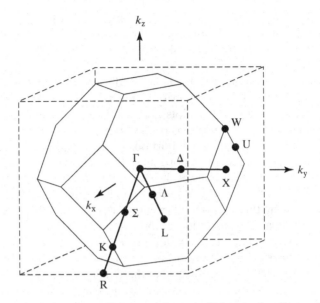

Figure 3.12   First Brillouin zone for materials crystallizing in the diamond and zincblende lattices. (After Blakemore.[1] Reprinted with permission.)

Γ... identifies the zone center ($k = 0$)

X... denotes the zone end along a $\langle 100 \rangle$ direction, and

L... denotes the zone end along a $\langle 111 \rangle$ direction.

Also note from Fig. 3.12 that the maximum magnitude of k varies with direction. In particular, $\Gamma \rightarrow L$, the length from the zone center to the zone boundary along a $\langle 111 \rangle$ direction, is $\sqrt{3}/2 \simeq 0.87$ times $\Gamma \rightarrow X$, the distance from the zone center to the zone boundary along a $\langle 100 \rangle$ direction. This should serve to explain the different widths of the $E-k\langle 100 \rangle$ and $E-k\langle 111 \rangle$ diagrams to be considered next.

3.3.2   *E-k* Diagrams

The presentation of *E-k* information characterizing 3-D crystals poses a fundamental problem. Since three dimensions are required to represent the *k*-vector, real-material *E-k* plots are intrinsically four-dimensional. A plot totally characterizing the band structure of a 3-D lattice is obviously impossible to construct. Constructing reduced-dimension plots where one or more of the variables is held constant is a possible solution, but the random production of such plots could become quite laborious. Fortunately, in semiconductor work only those portions of the bands normally occupied by carriers—the near vicinity of the conduction-band minima and the valence-band maxima—are routinely of interest. In the case of the diamond and zincblende lattices, the extrema points invariably occur at the zone center or lie along the high-symmetry $\langle 100 \rangle$ and $\langle 111 \rangle$ directions. Consequently, the information of greatest interest can be derived from plots of allowed energy versus the magnitude of k along these high-symmetry directions.

Figure 3.13 displays $\langle 100 \rangle / \langle 111 \rangle E\text{-}k$ diagrams characterizing the band structures in Ge (3.13a), Si (3.13b), and GaAs (3.13c and d). Before examining these figures it should be explained that the plots are two-direction composite diagrams. Because of crystal symmetry, the $-k$ portions of the $\langle 100 \rangle$ and $\langle 111 \rangle$ diagrams are just the mirror images of the corresponding $+k$ portions of the diagrams: no new information is conveyed by including the negative portions of the diagrams. It is therefore standard practice to delete the negative portions of the diagrams, turn the $\langle 111 \rangle$ diagrams so that the $+k$ direction faces to the left, and abut the two diagrams at $k = 0$. The respective positioning of L, denoting the zone boundary along a $\langle 111 \rangle$ direction, and X, denoting the zone boundary along a $\langle 100 \rangle$ direction, at the left- and right-hand ends of the k-axis corroborates the above observation. Likewise, the left-hand portions ($\Gamma \to L$) of the diagrams are shorter than the right-hand portions ($\Gamma \to X$) as expected from Brillouin-zone considerations. Also note that the energy scale in these diagrams is referenced to the energy at the top of the valence band. E_v is the maximum attainable valence-band energy, E_c the minimum attainable conduction-band energy, and $E_G = E_c - E_v$ the band gap energy.

Let us now examine the diagrams for factual information. One observes the following:

VALENCE BAND

(1) In all cases the valence-band maximum occurs at the zone center, at $k = 0$.

(2) The valence band in each of the materials is actually composed of three subbands. Two of the bands are degenerate (have the same allowed energy) at $k = 0$, while the third band maximizes at a slightly reduced energy. (In Si the upper two bands are indistinguishable on the gross energy scale used in constructing Fig. 3.13(b). Likewise, the maximum of the third band is a barely discernible 0.044 eV below E_v at $k = 0$.) Consistent with the effective-mass/energy-band-curvature discussions presented in Subsection 3.2.4, the $k = 0$ degenerate band with the smaller curvature about $k = 0$ is called the *heavy-hole* band, and the $k = 0$ degenerate band with the larger curvature is called the *light-hole* band. The subband maximizing at a slightly reduced energy is the *split-off* band.

(3) Near $k = 0$ the shape and therefore curvature of the subbands is essentially orientation independent. The significance of this observation will be explained later.

CONDUCTION BAND

(1) The gross features of the Ge, Si, and GaAs conduction-band structures are again somewhat similar. The conduction band in each case is composed of a number of subbands. The various subbands in turn exhibit localized and absolute minima at the zone center or along one of the high-symmetry directions. However—and this is very significant—the position of the overall conduction-band minimum, the "valley" where the electrons tend to congregate, varies from material to material.

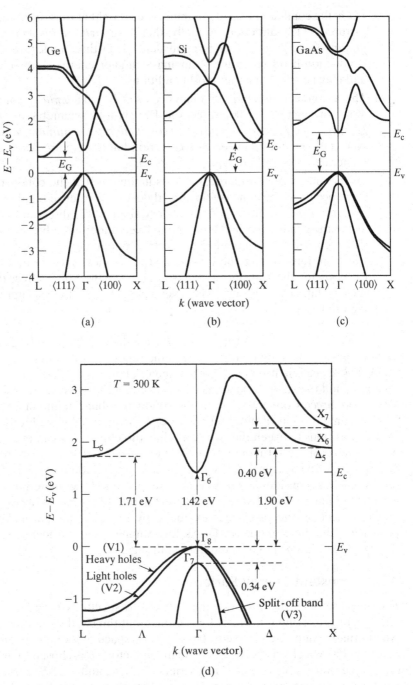

Figure 3.13 $\langle 100 \rangle / \langle 111 \rangle$ $E\text{-}k$ diagrams characterizing the conduction and valence bands of (a) Ge, (b) Si, and (c, d) GaAs. [(a–c) after Sze[2]; (d) from Blakemore.[1] Reprinted with permission.]

(2) In Ge the conduction-band minimum occurs right at the zone boundary along the pictured $\langle 111 \rangle$ direction. Actually, there are *eight equivalent conduction-band minima* since there are eight equivalent $\langle 111 \rangle$ directions. Other minima in the conduction-band structure occurring at higher energies are seldom populated and may be ignored under most conditions.

(3) The Si conduction-band minimum occurs at $k \simeq 0.8(2\pi/a)$ from the zone center along the pictured $\langle 100 \rangle$ direction. The six-fold symmetry of $\langle 100 \rangle$ directions gives rise of course to *six equivalent conduction-band minima* within the Brillouin zone. Other minima in the Si conduction-band structure occur at considerably higher energies and are typically ignored.

(4) Of the materials considered, GaAs is unique in that the conduction-band minimum occurs at the zone center directly over the valence-band maximum. Moreover, the L-valley at the zone boundary along $\langle 111 \rangle$ directions lies only 0.29 eV above the conduction-band minimum. Even under equilibrium conditions the L-valley contains a non-negligible electron population at elevated temperatures. Electron transfer from the Γ-valley to the L-valley provides the phenomenological basis for the transferred-electron devices (the Gunn-effect diode, etc.) and must be taken into account whenever a large electric field is impressed across the material.

Having discussed the properties of the conduction-band and valence-band structures separately, we should point out that the relative positioning of the band extrema points in k-space is in itself an important material property. When the conduction-band minimum and the valence-band maximum occur at the same value of k the material is said to be *direct*. Conversely, when the conduction-band minimum and the valence-band maximum occur at different values of k the material is said to be *indirect*. Electronic transitions between the two bands in a direct material can take place with little or no change in crystal momentum. On the other hand, conservation of momentum during an interband transition is a major concern in indirect materials. Of the three semiconductors considered, GaAs is an example of a direct material, while Ge and Si are indirect materials. The direct or indirect nature of a semiconductor has a very significant effect on the properties exhibited by the material, particularly the optical properties. The direct nature of GaAs, for example, makes it ideally suited for use in semiconductor lasers and infrared light-emitting diodes.

3.3.3 Constant-Energy Surfaces

E-k diagrams with the wavevector restricted to specific k-space directions provide one way of conveying relevant information about the band structures of 3-D crystals. An alternative approach is to construct a 3-D k-space plot of all the allowed k-values associated with a given energy E. For semiconductors, E is chosen to lie within the energy ranges normally populated by carriers ($E \lesssim E_v$ and $E \gtrsim E_c$). The allowed k-values form a surface or surfaces in k-space. The geometrical shapes, being associated with a given energy, are called *constant-energy surfaces*. Figure 3.14(a–c) displays the constant-energy surfaces characterizing the conduction-band structures near E_c in Ge, Si, and GaAs, respectively.

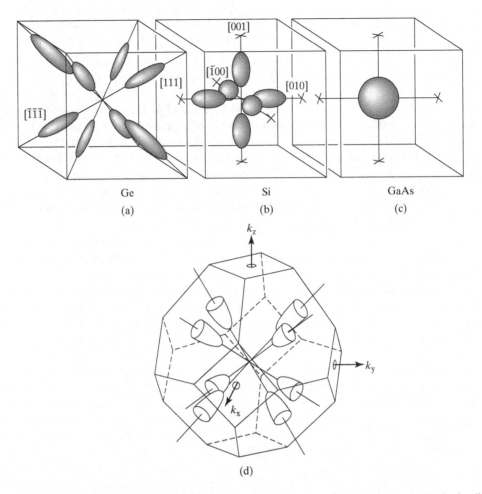

Figure 3.14   Constant-energy surfaces characterizing the conduction-band structure in (a, d) Ge, (b) Si, and (c) GaAs. (d) Shows the truncation of the Ge surfaces at the Brillouin-zone boundaries. [(a–c) after Sze[2] and Ziman[3]; (d) from McKelvey.[4] Reprinted with permission; the latter from Robert E. Krieger Publishing Co., Malabar, FL.]

As pointed out in the E-k diagram discussion, a Ge conduction-band minimum $(E = E_c)$ occurs along each of the eight equivalent $\langle 111 \rangle$ directions; a Si conduction-band minimum, along each of the six equivalent $\langle 100 \rangle$ directions. This explains the numbers and positions of the surfaces in the Ge and Si plots. (As clarified in Fig. 3.14(d), however, only one-half of the Ge surfaces actually lie within the first Brillouin zone—the Ge conduction-band minima occur right at the zone boundary.) The GaAs conduction-band minimum is of course positioned at the zone center, giving rise to a single constant-energy surface.

Although the preceding discussion covers most plot features, the geometrical shapes of the constant-energy surfaces still need to be explained. Working toward this end we recall that the E-k relationship characterizing one-dimensional crystals is approximately parabolic in the vicinity of the band extrema points. The three-dimensional case is completely analogous: for energies only slightly removed from E_c one can write in general

$$E - E_c \simeq Ak_1^2 + Bk_2^2 + Ck_3^2 \qquad (3.34)$$

where A, B, and C are constants and k_1, k_2, and k_3 are k-space coordinates measured from the center of a band minimum along principal axes. (Using the [111] Ge minimum of Fig. 3.14(a) as an example, the k_1-k_2-k_3 coordinate system would be centered at the [111] L-point and one of the coordinate axes, say the k_1-axis, would be directed along the k_x-k_y-k_z [111] direction.) For cubic crystals such as Ge, Si, and GaAs, at least two of the constants in Eq. (3.34) must be equal to satisfy symmetry requirements. Thus, for energies near the conduction-band minima in these materials, the allowed E-k relationships are

$$E - E_c \simeq A(k_1^2 + k_2^2 + k_3^2) \qquad (A = B = C) \qquad (3.35)$$

and (writing down only one of the three possible variants)

$$E - E_c \simeq Ak_1^2 + B(k_2^2 + k_3^2) \qquad (B = C) \qquad (3.36)$$

With E constant, Eq. (3.35) is readily recognized as the equation for a sphere centered at the band minimum. Eq. (3.36) with E held constant, on the other hand, is the mathematical expression for an ellipsoid of revolution, with k_1 being the axis of revolution. The GaAs conduction-band structure exhibits approximately spherical constant-energy surfaces described by Eq. (3.35)[†]; the Ge and Si constant-energy surfaces are all ellipsoids of revolution described by Eq. (3.36).

Examining the valence-band structure, one finds that the three subbands in Ge, Si, and GaAs are each characterized to first order by a plot identical to Fig. 3.14(c). In other words, the constant-energy surfaces about the $k = 0$ valence-band maxima are approximately spherical and are described to first order by Eq. (3.35) with $E - E_c \rightarrow E_v - E$. This is consistent with the earlier E-k diagram observation concerning the orientation independence of these subbands.[‡]

[†]For a more precise description of the GaAs conduction-band structure the reader is referred to Blakemore,[1] pp. R157-R160.

[‡]The interaction between the heavy- and light-hole subbands gives rise to an E-k perturbation which must be taken into account in more exacting computations. Including the perturbation one finds[5]

$$E_v - E = Ak^2 \pm [B^2k^4 + C^2(k_x^2k_y^2 + k_x^2k_z^2 + k_y^2k_z^2)]^{1/2}$$

with $k^2 = k_x^2 + k_y^2 + k_z^2$. Where the $(\pm)$ appears in the above equation, the $(+)$ is used in treating the light-hole band and the $(-)$ in treating the heavy-hole band. To be precise, therefore, the constant-energy surfaces about the $k = 0$ valence-band maximum are actually somewhat distorted spheres. A visualization of the Si heavy-hole band distortion can be found in Ziman,[3] p. 119.

Having investigated the construction and interpretation of constant-energy plots, one might wonder about the general utility of the plots. First and foremost, the constant-energy plots are very helpful visual and conceptual aids. From these plots one can ascertain by inspection the positions and multiplicity of band extrema points. As we will see, the shapes of the surfaces also provide information about the carrier effective masses. References to the plots are often encountered in advanced device analyses, particularly those involving orientation-dependent phenomena. Herein we will make specific use of the Fig. 3.14 plots during the effective mass discussion and in the subsequent density-of-states derivation (Subsection 4.1.2).

3.3.4 Effective Mass

General Considerations

In the one-dimensional analysis the electron motion resulting from an impressed external force was found to obey a modified form of Newton's second law, $dv/dt = F/m^*$. The scalar parameter, $m^* = \hbar^2/(d^2E/dk^2)$, was identified as the electron effective mass. In three-dimensional crystals the electron acceleration arising from an applied force is analogously given by

$$\frac{d\mathbf{v}}{dt} = \frac{1}{\mathbf{m}^*} \cdot \mathbf{F} \tag{3.37}$$

where

$$\frac{1}{\mathbf{m}^*} = \begin{pmatrix} m_{xx}^{-1} & m_{xy}^{-1} & m_{xz}^{-1} \\ m_{yx}^{-1} & m_{yy}^{-1} & m_{yz}^{-1} \\ m_{zx}^{-1} & m_{zy}^{-1} & m_{zz}^{-1} \end{pmatrix} \tag{3.38}$$

is the inverse effective mass tensor with components

$$\boxed{\frac{1}{m_{ij}} = \frac{1}{\hbar^2}\frac{\partial^2 E}{\partial k_i \partial k_j} \qquad \dots i, j = x, y, z} \tag{3.39}$$

An interesting consequence of the 3-D equation of motion is that the acceleration of a given electron and the applied force will not be colinear in direction as a general rule. For example, given a force pointing in the $+x$ direction, one obtains

$$\frac{d\mathbf{v}}{dt} = m_{xx}^{-1}F_x\mathbf{a}_x + m_{yx}^{-1}F_x\mathbf{a}_y + m_{zx}^{-1}F_x\mathbf{a}_z \tag{3.40}$$

with $\mathbf{a}_x$, $\mathbf{a}_y$, and $\mathbf{a}_z$ being unit vectors directed along the x, y, and z axes, respectively. Fortunately, the crystal and therefore the k-space coordinate system can always be rotated so as to align the k-space axes parallel to the principal axis system centered at a band extrema point. Since the E-k relationship is parabolic about the band extrema point, all $1/m_{ij}$ where $i \neq j$ will then vanish, thereby eliminating off-diagonal terms in the effective-mass tensor. In general, therefore, a maximum of three effective-mass components are needed to specify the motion of carriers energetically confined to the vicinity of extrema points. Moreover, the equation of motion in the rotated coordinate system drastically simplifies to $dv_i/dt = F_i/m_{ii}$.

Ge, Si, and GaAs

An even further simplification results when one considers cubic crystals such as Ge, Si, and GaAs. For GaAs, the k_x-k_y-k_z coordinate system is a principal-axes system and the conduction-band structure is characterized to first order by the "spherical" E-k relationship

$$E - E_c = A(k_x^2 + k_y^2 + k_z^2) \tag{3.41}$$

Thus, not only do the $1/m_{ij}$ components with $i \neq j$ vanish, but

$$m_{xx}^{-1} = m_{yy}^{-1} = m_{zz}^{-1} = \frac{2A}{\hbar^2} \tag{3.42}$$

Defining $m_{ii} = m_e^*$, we can therefore write

$$\boxed{E - E_c = \frac{\hbar^2}{2m_e^*}(k_x^2 + k_y^2 + k_z^2) \qquad \ldots \text{GaAs}} \tag{3.43}$$

and

$$\frac{d\mathbf{v}}{dt} = \frac{\mathbf{F}}{m_e^*} \tag{3.44}$$

For the conduction-band electrons in GaAs, the effective mass tensor reduces to a simple scalar, giving rise to an orientation-independent equation of motion like that of a classical particle. Obviously, spherical energy bands are the simplest type of band structure, necessitating a single effective-mass value for carrier characterization.

The characterization of conduction-band electrons in Ge and Si is only slightly more involved. For any of the ellipsoidal energy surfaces encountered in these materials one can always set up a k_1-k_2-k_3 principal-axis system where k_1 lies along the axis of revolution (see Fig. 3.15). The constant-energy surfaces are then described [repeating Eq. (3.36)] by

$$E - E_c = Ak_1^2 + B(k_2^2 + k_3^2) \tag{3.45}$$

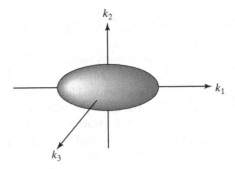

Figure 3.15   Principal-axis system for the ellipsoidal constant-energy surfaces in Ge and Si.

In the principal axis system the $i \neq j$ $(i, j = 1, 2, 3)$ components of the inverse effective-mass tensor are again zero and

$$m_{11}^{-1} = \frac{2A}{\hbar^2} \tag{3.46a}$$

$$m_{22}^{-1} = m_{33}^{-1} = \frac{2B}{\hbar^2} \tag{3.46b}$$

Because m_{11} is associated with the k-space direction lying along the axis of revolution, it is called the *longitudinal* effective mass and is usually given the symbol m_ℓ^*. Similarly, $m_{22} = m_{33}$, being associated with a direction perpendicular to the axis of revolution, is called the *transverse* effective mass and is given the symbol m_t^*. In terms of the newly introduced symbols we can write

$$E - E_c = \frac{\hbar^2}{2m_\ell^*}k_1^2 + \frac{\hbar^2}{2m_t^*}(k_2^2 + k_3^2) \qquad \ldots \text{Ge, Si} \tag{3.47}$$

Now, Eq. (3.47) models any of the ellipsoidal energy surfaces in Ge and Si. For a given material, however, all the ellipsoids of revolution have precisely the same shape. Consequently, the two effective-mass parameters, m_ℓ^* and m_t^*, totally characterize the conduction-band electrons in Ge and Si.

The relative sizes of m_ℓ^* and m_t^*, it should be noted, can be deduced by inspection from the Si and Ge constant-energy plots. By comparing Eq. (3.47) with the general expression for an ellipsoid of revolution, one finds

$$\frac{m_\ell^*}{m_t^*} = \left(\frac{\text{Length of the ellipsoid along the axis of revolution}}{\text{Maximum width of the ellipsoid perpendicular to the axis of revolution}} \right)^2 \tag{3.48}$$

Examining Fig. 3.14(a) and (b) we therefore conclude $m_\ell^* > m_t^*$ for both Ge and Si. The greater elongation of the Ge ellipsoids further indicates that the m_ℓ^*/m_t^* ratio is greater for Ge than for Si.

Finally, let us turn to the characterization of holes in Ge, Si, and GaAs. As established previously, the valence-band structure in these materials is approximately spherical and composed of three subbands. Thus, the holes in a given subband can be characterized by a single effective-mass parameter, but three effective masses are technically required to characterize the entire hole population. (The split-off band, being depressed in energy, is only sparsely populated and is often ignored.) The subband parameters are m_{hh}^*, the heavy hole effective mass; $m_{\ell h}^*$, the light hole effective mass; and m_{so}^*, the effective mass of holes in the split-off band.

Measurement

All of the effective masses introduced in the preceding discussion (m_e^*, m_ℓ^*, m_t^*, m_{hh}^*, $m_{\ell h}^*$, and m_{so}^*) are directly measurable material parameters. The parametric values have been obtained in a relatively straightforward manner from cyclotron resonance experiments. The near-extrema point band structure, multiplicity and orientation of band minima, etc. were, in fact, all originally confirmed by cyclotron resonance data.

In the basic cyclotron resonance experiment, the test material is situated in a microwave resonance cavity and cooled to liquid helium temperatures ($T \simeq 4$ K). A static magnetic field B and an rf electric field $\mathscr{E}$ oriented normal to the B-field are applied across the sample; the Q-factor of the resonant cavity, reflecting the absorption of rf $\mathscr{E}$-field energy, is monitored as a function of B-field strength.

The force exerted by the B-field causes the carriers in the sample to move in an orbit-like path about the direction of the B-field (see Fig. 3.16). The frequency of the orbit, called the cyclotron frequency ω_c, is directly proportional to the B-field strength and inversely dependent on the effective mass (or masses) characterizing the carrier orbit. When the B-field strength is adjusted such that ω_c equals the ω of the rf electric field, the carriers absorb energy from the electric field and a resonance, a peak in the Q-factor, is observed. From the B-field strength, the B-field orientation, and the ω at resonance, one can deduce the effective mass or the effective mass combination corresponding

B

rf $\mathscr{E}$

Figure 3.16 Carrier orbit and applied field orientations in the cyclotron resonance experiment.

Table 3.1 Electron and Hole Effective Masses in Ge,[6] Si[7], and GaAs[1] at 4 K. (All values referenced to the free electron rest mass m_0.)

Effective Mass	Ge	Si	GaAs
m_ℓ^*/m_0	1.588	0.9163	—
m_t^*/m_0	0.08152	0.1905	—
m_e^*/m_0	—	—	0.067[†]
m_{hh}^*/m_0	0.347	0.537	0.51
$m_{\ell h}^*/m_0$	0.0429	0.153	0.082
m_{so}^*/m_0	0.077	0.234	0.154

[†]Band edge effective mass. The E-k relationship about the GaAs conduction-band minimum becomes non-parabolic and m_e^* increases at energies only slightly removed from E_c.

to a given experimental configuration. Repeating the experiment for different B-field orientations allows one to separate out the effective mass factors (deduce both m_ℓ^* and m_t^*, for example) and to ascertain the orientational dependencies. The experiment is performed at low temperatures to maximize the number of orbits completed by the carriers between scattering events. Orbit disruption due to carrier scattering increases with temperature and tends to broaden and eventually eliminate resonance peaks. (For more information about the cyclotron resonance experiment, the reader is referred to ref. [5] for experimental details and to ref. [4] for an extended explanatory discussion.)

The 4 K effective mass values deduced from cyclotron resonance experiments for Ge, Si, and GaAs are listed in Table 3.1. The entries in this table were derived from definitive review-type works by Paige[6] on Ge, by Barber[7] on Si, and by Blakemore[1] on GaAs. Note that, as previously inferred from the Fig. 3.14 constant-energy plots, $m_\ell^* > m_t^*$ (Ge, Si) and m_ℓ^*/m_t^* (Ge) $> m_\ell^*/m_t^*$ (Si).

In practical computations one would often like to know the temperature dependence of the effective mass parameters and, in particular, the parametric values at room temperature. Unfortunately, cyclotron resonance experiments cannot be performed at room temperature and limited data is available concerning the temperature dependence of these parameters. A theoretical extrapolation of the effective mass values to and above room temperature has nevertheless been performed by Barber[7] for Si and by Blakemore[1] for GaAs. As a matter of expediency, the effective masses are often implicitly assumed to be temperature independent and the 4 K values simply employed for all operational temperatures. In most cases the errors thereby introduced appear to be relatively minor.

3.3.5 Band Gap Energy

The band gap energy, $E_G = E_c - E_v$, is perhaps the most important parameter in semiconductor physics. At room temperature (roughly $T = 300$ K), $E_G(\text{Ge}) \simeq 0.66$ eV,

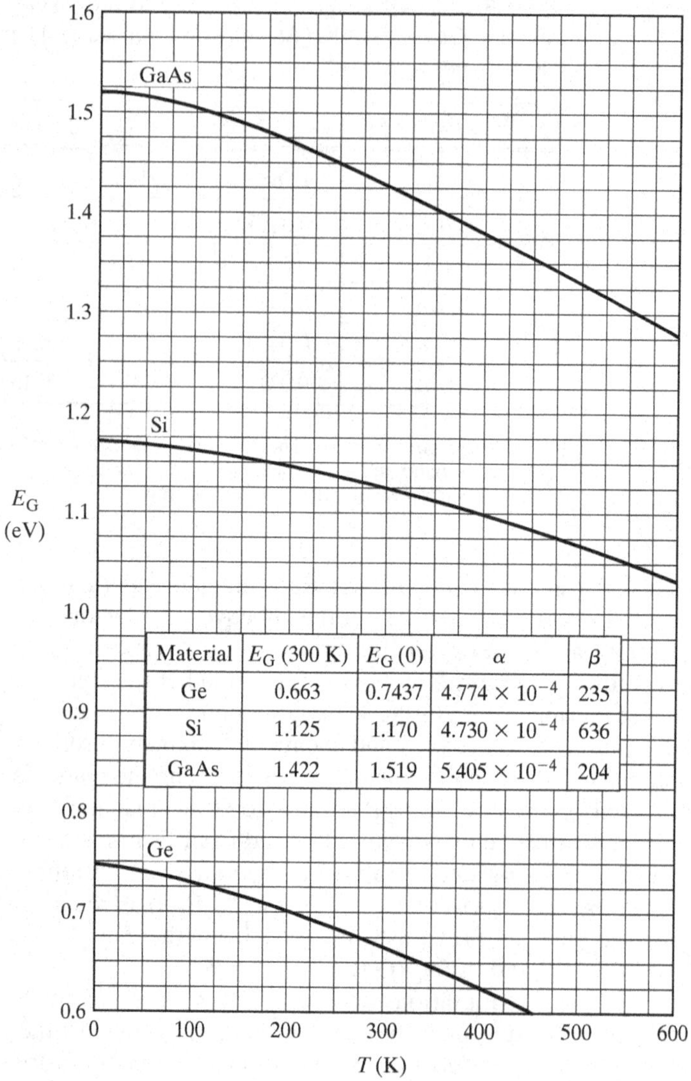

Material	E_G (300 K)	E_G (0)	α	β
Ge	0.663	0.7437	4.774×10^{-4}	235
Si	1.125	1.170	4.730×10^{-4}	636
GaAs	1.422	1.519	5.405×10^{-4}	204

Figure 3.17 The band gap energy in Ge, Si, and GaAs as a function of temperature. The insert gives the 300 K values and the Eq. (3.49) fit parameters. (After Sze.[2] Reprinted with permission.)

$E_G(\text{Si}) \simeq 1.12$ eV, and $E_G(\text{GaAs}) \simeq 1.42$ eV. With decreasing temperature a contraction of the crystal lattice usually leads to a strengthening of the interatomic bonds and an associated increase in the band gap energy. This is true for the vast majority of semiconductors including Ge, Si, and GaAs. To a very good approximation, the cited variation of band gap energy with temperature can be modeled by the "universal" empirical relationship

$$E_G(T) = E_G(0) - \frac{\alpha T^2}{(T + \beta)} \tag{3.49}$$

where α and β are constants chosen to obtain the best fit to experimental data and $E_G(0)$ is the limiting value of the band gap at zero Kelvin. The band gap versus temperature dependencies deduced from Eq. (3.49) for Ge, Si, and GaAs are plotted in Fig. 3.17; the fit parameters are specified in the figure insert. A more complete tabulation of semiconductor band gaps and other pertinent E-k information can be found in Appendix A of Wolfe et al.[8]

REFERENCES

[1] J. S. Blakemore, "Semiconducting and Other Major Properties of Gallium Arsenide," J. Appl. Phys., *53*, R123 (Oct., 1982).

[2] S. M. Sze, *Physics of Semiconductor Devices*, 2nd edition, John Wiley & Sons, Inc., New York, 1981.

[3] J. M. Ziman, *Electrons and Phonons, The Theory of Transport Phenomena in Solids*, Oxford University Press, London, 1960.

[4] J. P. McKelvey, *Solid State and Semiconductor Physics*, Harper and Row, New York, 1966.

[5] G. Dresselhaus, A. F. Kip, and C. Kittel, "Cyclotron Resonance of Electrons and Holes in Silicon and Germanium," Phys. Rev., *98*, 368 (1955).

[6] E. G. S. Paige, *The Electrical Conductivity of Germanium (Progress in Semiconductors*, Vol. 8, edited by A. F. Gibson and R. E. Burgess), John Wiley & Sons, Inc., New York, 1964.

[7] H. D. Barber, "Effective Mass and Intrinsic Concentration in Silicon," Solid-State Electronics, *10*, 1039 (1967).

[8] C. M. Wolfe, N. Holonyak Jr., and G. E. Stillman, *Physical Properties of Semiconductors*, Prentice Hall, Englewood Cliffs, NJ, 1989.

PROBLEMS

3.1 Answer the following questions as concisely as possible.

  (a) State the Bloch theorem for a one-dimensional system.

  (b) The current associated with the motion of electrons in a totally filled energy band (a band in which all allowed states are occupied) is always identically zero. Briefly explain why.

  (c) Define in words what is meant by a "Brillouin zone."

  (d) Because of crystal symmetry one would expect the one-dimensional E versus k plots characterizing cubic crystals to be symmetrical about the Γ point. Why aren't the E versus k plots in Fig. 3.13 symmetrical about the Γ point?

3.2 The Kronig–Penney model solution presented in Subsection 3.2.2 is somewhat unconventional. To indicate why a nonstandard solution approach was presented (and, more generally, to coerce the reader into examining the Kronig-Penney model solution), let us review the usual textbook approach. (You may wish to consult reference [4], pp. 213–214, to check your answers.)

(a) From the Bloch theorem we know

$$\psi(x) = e^{ikx}u(x)$$

where $u(x)$ is the unit cell wavefunction. Substitute the above expression for $\psi(x)$ into Schrödinger's equation and obtain the simplest possible differential equation for $u_a(x)$ [$u(x)$ for $0 < x < a$] and $u_b(x)$ [$u(x)$ for $-b < x < 0$]. Introduce α and β as respectively defined by Eqs. (3.8) and (3.10).

(b) Record the general solutions for $u_a(x)$ and $u_b(x)$.

(c) Indicate the boundary conditions on the unit cell wavefunctions.

(d) Apply the part (c) boundary conditions to obtain a set of four simultaneous equations analogous to Eqs. (3.13).

(e) Show that Eq. (3.15) again results when one seeks a nontrivial solution to the part (d) set of equations. (NOTE: This part of the problem involves a considerable amount of straightforward but very tedious algebra.)

3.3 The comparison between the free-particle solution and the extended-zone representation of the Kronig–Penney model solution (Fig. 3.6) can be found in a number of texts. In the majority of texts the free-particle solution is pictured to be coincident with the lower energy band of the Kronig–Penney solution at all zone boundaries. Verify that Fig. 3.6 was constructed properly; i.e., verify that the dashed line, the free-particle solution, was drawn through the correct points in the figure.

3.4 A certain hypothetical material with cubic crystal symmetry is characterized by the E-k plot sketched in Fig. P3.4.

(a) Which set of holes, band A holes or band B holes, will exhibit the greater [100]-direction (m_{xx}) effective mass? Explain.

(b) Sketch the expected form of the valence-band constant-energy surfaces for the represented cubic material. Assume that the E-k relationship is parabolic (i.e., an ellipsoid of revolution)

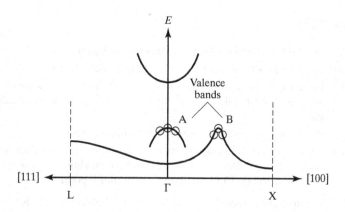

Figure P3.4

in the vicinity of the B maxima. Include the constant energy surfaces associated with both the A and B bands.

3.5 The E-k relationship characterizing an electron confined to a *two-dimensional* surface layer is of the form

$$E - E_c = \frac{\hbar^2 k_x^2}{2m_1} + \frac{\hbar^2 k_y^2}{2m_2} \quad \dots m_1 \neq m_2$$

An electric field is applied in the x–y plane at a 45 degree angle to the x–axis. Taking the electron to be initially at rest, determine its direction of motion in the x–y plane.

3.6 Consider the detailed E-k diagram for GaAs reproduced in text Fig. 3.13(d).

(a) How does one deduce the 300 K value of E_G from the diagram. Is the value of $E_G(300\,\text{K})$ deduced from the diagram consistent with that quoted in the Fig. 3.17 insert?

(b) How does one deduce that GaAs is a direct semiconductor from the given E-k diagram?

(c) As best as can be determined from Fig. 3.13(d), is the pictured E-k diagram consistent with the effective masses for GaAs quoted in Table 3.1? Explain.

3.7 (a) The E-k relationship about the GaAs conduction-band minimum becomes non-parabolic at energies only slightly removed from E_c and is more accurately described by

$$E - E_c = ak^2 - bk^4 \quad (a > 0, b > 0)$$

What effect will the cited fact have on the effective mass of electrons in the GaAs conduction band? Substantiate your conclusion. (Is your answer here in agreement with the Table 3.1 footnote?)

(b) Electrons in GaAs can transfer from the Γ minimum to the L minima at sufficiently high electric fields. If electrons were to transfer from the Γ minimum to the L minimum shown in Fig. 3.13(d), would their effective mass increase or decrease? Explain. (The constant-energy surfaces about the L minima are actually ellipsoidal, but for simplicity assume the surfaces to be spherical in answering this question.)

3.8 Like GaAs, GaP crystallizes in the zincblende lattice and the valence band maxima occur at the Γ-point in the first Brillouin zone. Unlike GaAs, the conduction band minima in GaP occur at the X-points in the Brillouin zone.

(a) Where are the X-points located in k-space?

(b) Is GaP a direct or indirect material? Explain.

(c) Given that the constant energy surfaces at the X-points are ellipsoidal with $m_\ell^*/m_0 = 1.12$ and $m_t^*/m_0 = 0.22$, what is the ratio of the longitudinal length to the maximum transverse width of the surfaces?

(d) Picturing only that portion of the constant energy surfaces within the first Brillouin zone, construct a constant-energy surface diagram characterizing the conduction-band structure in GaP.

3.9 (a)   Derive Eq. (3.48).

HINT: Show that Eq. (3.47) can be manipulated into the form

$$\frac{k_1^2}{\alpha^2} + \frac{k_2^2 + k_3^2}{\beta^2} = 1$$

where

$$\alpha \equiv \sqrt{(2m_\ell^*/\hbar^2)(E - E_c)}$$
$$\beta \equiv \sqrt{(2m_t^*/\hbar^2)(E - E_c)}$$

Confirm (quote your reference) that the above k_1-k_2-k_3 expression is the defining equation for a prolate spheroid—the ellipsoidal surface formed by the rotation of an ellipse about its major axis. 2α and 2β are, respectively, the lengths of the major and minor axes of the rotated ellipse.

(b)   Are the Fig. 3.14 ellipsoidal surfaces for Ge and Si in general agreement with the m_ℓ^*/m_t^* ratios deduced from Table 3.1? Explain.

3.10   In Table 3 on p. 1318 of R. Pässler, Solid State Electronics, *39*, 1311 (1996), there is a listing of what the author considers to be the most accurate silicon E_G versus temperature data and the corresponding computed E_G values from "superior" empirical fits to the experimental data. Using Eq. (3.49) and the parameters listed in the text Fig. 3.17 insert, compute E_G to 5 places at 50 K intervals from 50 K to 500 K. Compare your computed E_G values with the tabulated values in the cited reference.

CHAPTER 4

Equilibrium Carrier Statistics

The energy band considerations of Chapter 3 provided fundamental information about the carriers inside a semiconductor. Two types of carriers, $-q$ charged electrons in the conduction band and $+q$ charged holes in the valence band, were identified as contributing to charge transport. The allowed energies and crystal momentum available to the carriers in the respective bands were specified. It was also established that the motion of carriers in a crystal can be visualized and described in a quasi-classical manner using the effective mass approximation. Device modeling, however, requires additional information of a statistical nature. The desired information includes, for example, the precise number of carriers in the respective bands and the energy distribution of carriers within the bands. In this chapter we develop the quantitative relationships that are used to characterize the carrier populations inside semiconductors under equilibrium conditions. Equilibrium is the condition that prevails if the semiconductor has been left unperturbed for an extended period of time. (A more exacting definition of the equilibrium state will be presented in Chapter 5.) Essentially all practical material and device computations use the equilibrium condition as a point of reference.

The first two sections of this chapter contain derivations of the well-known expressions for the density and occupation of energy band states. In lower-level texts these expressions—the density of states function and the Fermi function—are often presented without justification. The derivation of the functions is intrinsically worthwhile, however, in that it provides a deeper understanding of the general subject matter. The density of states derivation also establishes the tie between the density of states effective masses and the effective masses introduced in the preceding chapter. The Fermi function derivation pointedly underscores the almost universal applicability of the expression. Most importantly, though, knowledge of the standard derivation permits one to modify the result for situations that violate derivational assumptions. For example, the "two-dimensional" density of states encountered in the operational analysis of the MODFET (Modulation-Doped Field-Effect Transistor) can be established in a straightforward manner by modifying the standard density of states derivation. Likewise, the occupancy of states in the band gap is slightly different from the occupancy of band states. This difference can be readily understood if one is familiar with the derivation of the Fermi function.

The remaining sections in this chapter primarily deal with the development and use of carrier concentration relationships. Material routinely covered in introductory texts will be simply reviewed herein. Topics not often found in introductory texts will be explored in greater detail.

4.1 DENSITY OF STATES

The density of states is required as the first step in determining the carrier concentrations and energy distributions of carriers within a semiconductor. Integrating the density of states function, $g(E)$, between two energies E_1 and E_2 tells one the number of allowed states available to electrons in the cited energy range per unit volume of the crystal. In principle, the density of states could be determined from band theory calculations for a given material. Such calculations, however, would be rather involved and impractical. Fortunately, an excellent approximation for the density of states near the band edges, the region of the bands normally populated by carriers, can be obtained through a much simpler approach. As best visualized in terms of Fig. 4.1(a), which is a modified version of Fig. 3.4, electrons in the conduction band are essentially free to roam throughout the crystal. For electrons near the bottom of the band, the band itself forms a pseudo-potential well as pictured in Fig. 4.1(b). The well bottom lies at E_c and the termination of the band at the crystal surfaces forms the walls of the well. Since the energy of the electrons relative to E_c is typically small compared with the surface barriers, one effectively has a particle in a three-dimensional box. The density of states

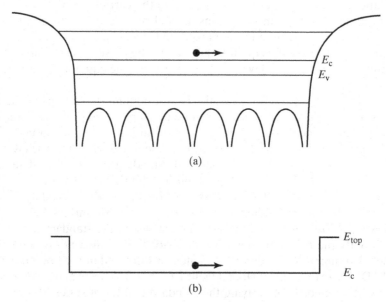

(a)

(b)

Figure 4.1 (a) Visualization of a conduction band electron moving in a crystal. (b) Idealized pseudo-potential well formed by the crystal surfaces and the band edges.

near the band edges can therefore be equated to the density of states available to a particle of mass m^* in a box with the dimensions of the crystal.

In the following we first work out the density of available states for a particle of mass m in a three-dimensional box. The result is subsequently modified to account for the carrier effective masses and the band structures of real materials.

4.1.1   General Derivation

Consider a particle of mass m and fixed total energy E confined to a crystal-sized box. As pictured in Fig. 4.2, the x, y, and z side-lengths of the box are taken to be a, b, and c, respectively. $U(x, y, z) = $ constant everywhere inside the box and is set equal to zero without any loss in generality. For the specified problem the time-independent Schrödinger equation then becomes

$$\frac{\partial^2 \psi}{\partial x^2} + \frac{\partial^2 \psi}{\partial y^2} + \frac{\partial^2 \psi}{\partial z^2} + k^2 \psi = 0 \qquad \dots \begin{array}{l} 0 < x < a \\ 0 < y < b \\ 0 < z < c \end{array} \qquad (4.1)$$

where

$$k \equiv \sqrt{2mE/\hbar^2} \quad \text{or} \quad E = \frac{\hbar^2 k^2}{2m} \qquad (4.2)$$

To solve Eq. (4.1) we employ the separation of variables technique. Specifically, the wavefunction solution is assumed to be of the form

$$\psi(x, y, z) = \psi_x(x)\psi_y(y)\psi_z(z) \qquad (4.3)$$

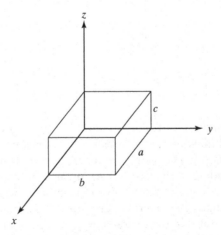

Figure 4.2 Envisioned crystal-sized box (infinitely deep three-dimensional potential well) with x, y, and z dimensions of a, b, and c, respectively.

Substituting Eq. (4.3) into Eq. (4.1) and dividing all terms by $\psi_x\psi_y\psi_z$, one obtains

$$\frac{1}{\psi_x}\frac{d^2\psi_x}{dx^2} + \frac{1}{\psi_y}\frac{d^2\psi_y}{dy^2} + \frac{1}{\psi_z}\frac{d^2\psi_z}{dz^2} + k^2 = 0 \qquad (4.4)$$

Since k^2 is a constant and the remaining terms in Eq. (4.4) are functions of a single spatial coordinate, the equation will be satisfied for all possible values of x, y, and z if and only if the terms involving ψ_x, ψ_y, and ψ_z are individually equal to a constant. Thus

$$\frac{1}{\psi_x}\frac{d^2\psi_x}{dx^2} = \text{constant} = -k_x^2 \qquad (4.5a)$$

or

$$\frac{d^2\psi_x}{dx^2} + k_x^2\psi_x = 0 \qquad \ldots 0 < x < a \qquad (4.5b)$$

The separation constant has been logically identified as $-k_x^2$ because of its position as the coefficient of the ψ_x term in the (4.5b) one-dimensional wave equation. Analogous equations involving k_y^2 and k_z^2 can be written for ψ_y and ψ_z. Clearly, the three-dimensional problem has been reduced to three one-dimensional problems. Referring to the particle-in-a-1D-box analysis presented in Subsection 2.3.2, we rapidly arrive at the overall wavefunction solution:

$$\psi(x,\ y,\ z) = A\ \sin k_x x\ \sin k_y y\ \sin k_z z \qquad (4.6)$$

$$k^2 = k_x^2 + k_y^2 + k_z^2 \qquad (4.7)$$

$$k_x = \frac{n_x\pi}{a}; \qquad k_y = \frac{n_y\pi}{b}; \qquad k_z = \frac{n_z\pi}{c} \qquad (4.8)$$

$$n_x,\ n_y,\ n_z = \pm1,\ \pm2,\ \pm3,\ \cdots \qquad (4.9)$$

The combination of Eqs. (4.2) and (4.7) to (4.9) yields the allowed particle energies. Our goal, it must be remembered, is to determine the number of allowed states as a function of energy. This would be a very simple task if we were only interested in states corresponding to the first few n_x, n_y, n_z combinations yielding energies within a specified energy range. However, given the large dimensions of a crystal-sized box, and the correspondingly small increments in the k's for unit changes in the n's, there could easily be $\sim10^{20}$ states in the overall energy range of interest near $E = 0$. Obviously, a more sophisticated counting technique must be employed.

To assist in the counting process, it is common practice to use the construct shown in Fig. 4.3(a). Each Schrödinger equation solution can be uniquely associated with a k-space vector, $\mathbf{k} = (\mathbf{n_x}\pi/a)\mathbf{a} + (\mathbf{n_y}\pi/b)\mathbf{b} + (\mathbf{n_z}\pi/c)\mathbf{c}$, where $\mathbf{a}$, $\mathbf{b}$, and $\mathbf{c}$ are unit vectors directed along the k-space coordinate axes. In the construct, the k-vector endpoints, each representing one Schrödinger equation solution, are recorded as dots on a three-dimensional k-space plot. A sufficient number of points are included to illustrate the general k-space periodicity of the solutions.[†]

Taking note of the lattice-like arrangement of the solution dots, one readily deduces that a k-space "unit cell" (see Fig. 4.3(b)) of volume $(\pi/a)(\pi/b)(\pi/c)$ contains one allowed solution, or

$$\left(\frac{\text{Solutions}}{\text{Unit volume of } k\text{-space}}\right) = \frac{abc}{\pi^3} \tag{4.10}$$

The solutions/unit k-volume and allowed states/unit k-volume, however, are not precisely equivalent. First, as pointed out in Subsection 2.3.2, there is no physical difference between wavefunction solutions which differ only in sign. Thus, for example, the $(\mathbf{n_x}, \mathbf{n_y}, \mathbf{n_z}) = (1, 1, 1)$ solution and the seven other solutions involving combinations of $\mathbf{n_x} = \pm 1$, $\mathbf{n_y} = \pm 1$, and $\mathbf{n_z} = \pm 1$ are actually one and the same allowed state. If one counts all the points on the k-space plot, it is therefore necessary to divide by eight to obtain the number of *independent solutions* per unit volume of k-space. (Although not

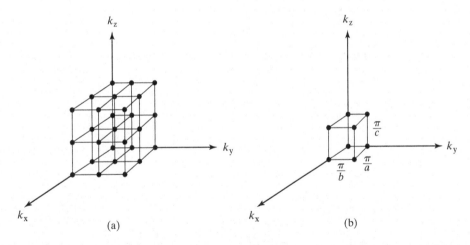

Figure 4.3   (a) k-space representation of Schrödinger equation solutions for a particle in a crystal-sized three-dimensional box. (b) k-space unit cell for solution points.

[†]Contrary to the situation pictured in Fig. 4.3(a), there are actually no allowed solutions lying on the coordinate planes; setting $\mathbf{n_x}$, $\mathbf{n_y}$, or $\mathbf{n_z} = 0$ yields the trivial $\psi = 0$ result. Because of the large number of states involved, however, the inclusion of the null-point values introduces an infinitesimally small error.

typically implemented, one could alternatively count only the solutions in the first k-space octant.) Secondly, the foregoing analysis completely neglected the quantum mechanical property known as spin. For electrons, two allowed spin states, spin up and spin down, must be associated with each independent solution. In summary, then, to obtain the allowed electron states per unit volume of k-space, the Eq. (4.10) result is divided by eight to eliminate redundant solutions and multiplied by two to account for spin. We therefore obtain

$$\left(\frac{\text{Allowed energy states}}{\text{Unit volume of } k\text{-space}} \right) = \frac{abc}{4\pi^3} \tag{4.11}$$

For reasons that will become obvious shortly, the next step is to determine the number of states with a k-value between an arbitrarily chosen k and $k + dk$. This is equivalent to adding up the states lying between the two k-space spheres pictured in Fig. 4.4. Again, owing to the large dimensions of the assumed real-space box and the corresponding close-packed density of k-space states (which places a large number of states within the spherical shell), the desired result is simply obtained by multiplying the k-space volume between the two spheres, $4\pi k^2 dk$, times the Eq. (4.11) expression for the allowed states per unit k-space volume; i.e.,

$$\left(\begin{array}{c} \text{Energy states with } k \\ \text{between } k \text{ and } k + dk \end{array} \right) = (4\pi k^2 dk)(abc/4\pi^3) \tag{4.12}$$

With the aid of Eq. (4.2), the just-determined states in an incremental dk range can be readily converted to the states in an incremental dE range. Specifically,

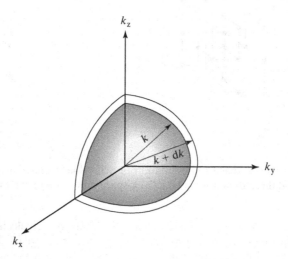

Figure 4.4 k-space spheres of radius k and $k + dk$ used to determine the allowed electronic states in an incremental dk range.

$$E = \hbar^2 k^2/2m \ldots \text{ or } \ldots k^2 = 2mE/\hbar^2 \tag{4.13}$$

$$dE = \hbar^2 k dk/m \ldots \text{ or } \ldots dk = \frac{1}{\hbar}\sqrt{\frac{m}{2}}\frac{dE}{\sqrt{E}} \tag{4.14}$$

and direct substitution into Eq. (4.12) yields

$$\begin{pmatrix} \text{Energy states with } E \\ \text{between } E \text{ and } E + dE \end{pmatrix} = (abc)\frac{m\sqrt{2mE}}{\pi^2\hbar^3}\,dE \tag{4.15}$$

Finally, by definition,

$$g(E) = \begin{pmatrix} \text{Energy states with } E \\ \text{between } E \text{ and } E + dE \end{pmatrix} \Big/ \mathcal{V}\,dE \tag{4.16}$$

where $\mathcal{V}$ is the volume of the crystal and $g(E)$ is the density of states. Thus, substituting Eq. (4.15) into Eq. (4.16) and noting that $\mathcal{V} = abc$, we arrive at the desired end result:

$$\boxed{g(E) = \frac{m\sqrt{2mE}}{\pi^2\hbar^3}} \tag{4.17}$$

4.1.2 Specific Materials

To obtain the conduction and valence band densities of states near the band edges in real materials, the mass m of the particle in the foregoing derivation [and in the Eq. (4.17) result] is replaced by the appropriate carrier effective mass. Also, if E_c is taken to be the minimum electron energy in the conduction band and E_v the maximum hole energy in the valence band, the E in Eq. (4.17) must be replaced by $E - E_c$ in treating conduction band states and by $E_v - E$ in treating valence band states. Introducing the subscripts c and v to identify the conduction and valence band densities of states, respectively, we can then write in general

$$\boxed{\begin{aligned} g_c(E) &= \frac{m_n^*\sqrt{2m_n^*(E - E_c)}}{\pi^2\hbar^3} \qquad \ldots E \geq E_c \tag{4.18a} \\[2ex] g_v(E) &= \frac{m_p^*\sqrt{2m_p^*(E_v - E)}}{\pi^2\hbar^3} \qquad \ldots E \leq E_v \tag{4.18b} \end{aligned}}$$

where m_n^* and m_p^* are the electron (n) and hole (p) *density of states effective masses.*

A new effective mass has been introduced in writing down the density of states expressions because we know that the band-structure description of carriers within a band often involves two or more effective masses. In general, therefore, the effective mass used in the density of states expression must be some combination, an appropriate "average," of the band-structure effective masses. The remaining task is to determine the precise relationship between the density of states effective masses and the band-structure effective masses.

Conduction Band—GaAs

As discussed in Chapter 3, the GaAs conduction band structure is approximately spherical and the electrons within the band are characterized by a single isotropic effective mass, m_e^*. For this special case the mass m of the particle-in-the-3D-box derivation can be directly replaced by m_e^*. Consequently,

$$\boxed{m_n^* = m_e^*} \qquad \ldots \text{GaAs} \qquad (4.19)$$

Conduction Band—Si, Ge

The conduction band structure in Si and Ge is characterized by ellipsoidal energy surfaces centered, respectively, at points along the $\langle 100 \rangle$ and $\langle 111 \rangle$ directions in k-space (see Fig. 3.14). In a k_1-k_2-k_3 principal-axes coordinate system, the mathematical description of the ellipsoidal energy surfaces was established to be

$$E - E_c = \frac{\hbar^2}{2m_\ell^*}k_1^2 + \frac{\hbar^2}{2m_t^*}(k_2^2 + k_3^2) \qquad (4.20)$$
$$\text{(Same as 3.47)}$$

which can be manipulated into the form

$$\frac{k_1^2}{\alpha^2} + \frac{k_2^2 + k_3^2}{\beta^2} = 1 \qquad (4.21)$$

$$\alpha \equiv \sqrt{\frac{2m_\ell^*(E - E_c)}{\hbar^2}} \qquad (4.22a)$$

$$\beta \equiv \sqrt{\frac{2m_t^*(E - E_c)}{\hbar^2}} \qquad (4.22b)$$

Equation (4.21) is the general form of the expression for a prolate spheroid—the ellipsoidal surface formed by the rotation of an ellipse about its major axis. 2α and 2β are, respectively, the lengths of the major and minor axes of the rotated ellipse. The volume of k-space enclosed by a single prolate spheroid is $(4/3)\pi\alpha\beta^2$.

If we next consider using an isotropic effective mass m_n^* in the particle-in-a-3D-box derivation, the inner k-space sphere of Fig. 4.4 employed in adding up allowed states would have a radius of $k_{eff} = \sqrt{2m_n^*(E - E_c)}/\hbar^2$. The volume of k-space enclosed by this "effective" constant-energy surface would be $(4/3)\pi k_{\text{eff}}^3$. If, for an arbitrarily chosen E, we now adjust m_n^* such that the k-space volumes, and hence the total number of states, enclosed by the actual (ellipsoidal) and effective (spherical) constant-energy surfaces are identical, the m_n^*-modified particle-in-a-box derivation will yield the correct density of conduction band states for Si and Ge. Equating the actual and effective k-space volumes, one obtains

$$N_{e\ell}\left(\frac{4}{3}\pi\alpha\beta^2\right) = \frac{4}{3}\pi k_{\text{eff}}^3 \tag{4.23}$$

or

$$N_{e\ell}(m_\ell^* m_t^{*2})^{1/2} = (m_n^*)^{3/2} \tag{4.24}$$

where $N_{e\ell}$ is the number of ellipsoidal surfaces lying within the first Brillouin zone. For Si, $N_{e\ell} = 6$; for Ge, $N_{e\ell} = (1/2)8 = 4$. (Only one-half of the eight Ge ellipsoids lie inside the first Brillouin zone.) We therefore conclude that

$$\boxed{m_n^* = 6^{2/3}(m_\ell^* m_t^{*2})^{1/3}} \quad \ldots \text{Si} \tag{4.25a}$$

$$\boxed{m_n^* = 4^{2/3}(m_\ell^* m_t^{*2})^{1/3}} \quad \ldots \text{Ge} \tag{4.25b}$$

Valence Band—Ge, Si, GaAs

Like the GaAs conduction band, the valence band structures of Ge, Si, and GaAs are all characterized by approximately spherical constant-energy surfaces. However, in each case there are two $k = 0$ degenerate subbands. (The split-off band, being depressed in energy, is typically ignored.) The subbands are respectively populated by heavy holes with an isotropic effective mass m_{hh}^* and light holes with an isotropic effective mass $m_{\ell\text{h}}^*$. Clearly, the particle-in-a-3D-box derivation can be separately applied to the light and heavy hole subbands. The total valence band density of states is then the sum of the two subband densities of states. We can therefore write

$$g_v(E) = \frac{m_p^*\sqrt{2m_p^*(E_v - E)}}{\pi^2\hbar^3} \tag{4.26a}$$

$$= \frac{m_{\text{hh}}^*\sqrt{2m_{\text{hh}}^*(E_v - E)}}{\pi^2\hbar^3} + \frac{m_{\ell\text{h}}^*\sqrt{2m_{\ell\text{h}}^*(E_v - E)}}{\pi^2\hbar^3} \tag{4.26b}$$

Table 4.1 Density of States Effective Masses for Ge, Si, and GaAs

Effective Mass		Ge	Si	GaAs
m_n^*/m_0	$T = 4$ K	0.553	1.062	0.067
	$T = 300$ K	....	1.182	0.0655[†]
m_p^*/m_0	$T = 4$ K	0.357	0.590	0.532
	$T = 300$ K	....	0.81	0.524

[†]The band edge effective mass ratio is 0.0632. The value quoted here takes into account the non-parabolic nature of the GaAs conduction band and yields the correct nondegenerate carrier concentrations when employed in computational expressions developed later in this chapter.

Thus

$$(m_p^*)^{3/2} = (m_{hh}^*)^{3/2} + (m_{\ell h}^*)^{3/2} \tag{4.27a}$$

or

$$\boxed{m_p^* = [(m_{hh}^*)^{3/2} + (m_{\ell h}^*)^{3/2}]^{2/3}} \tag{4.27b}$$

The density of states effective masses for Ge, Si, and GaAs are presented in Table 4.1. The 4 K entries in the table were calculated using the relationships developed in this subsection and the band-structure effective masses recorded in Table 3.1. The 300 K entries are theoretically extrapolated estimates taken from Barber[1] for Si and from Blakemore[2] for GaAs.

4.2 FERMI FUNCTION

Introduction

The Fermi function, $f(E)$, is a probability distribution function that tells one the ratio of filled to total allowed states at a given energy E. As we will see, statistical arguments are employed to establish the general form of the function. Basically, the electrons are viewed as indistinguishable "balls" that are being placed in allowed-state "boxes." Each box is assumed to accommodate a single ball. The boxes themselves are grouped into rows, the number of boxes per row corresponding to the allowed electronic states at a given energy. The numerical occurrence of all possible arrangements of balls per row yielding the same overall system energy is determined statistically, and the most likely arrangement identified. Finally, the Fermi function is equated to the most likely arrangement of balls (electrons) per row (energy). The cited arrangement, it turns out, occurs more often than all other arrangements combined. Moreover, the distribution of arrangements is highly peaked about the most probable arrangement. Thus it is reasonable to use the Fermi function—the most probable arrangement—to describe the filling of allowed states in actual electronic systems.

Problem Specification

We consider the placement of N electrons into a multi-level energy system. The assumed system is totally arbitrary: our considerations are not restricted to a specific material or set of materials. The system, as pictured in Fig. 4.5, contains S_i available states at an allowed energy E_i ($i = 1, 2, 3, \cdots$). N_i is taken to be the number of electrons with energy E_i. The electrons are assumed to be indistinguishable: the interchange of any two electrons would leave the electronic configuration unperturbed. Also, the likelihood of filling an individual state is taken to be energy independent.

The placement of electrons in the various allowed states is subject, however, to the following restrictions:

(1) Each allowed state can accommodate one and only one electron.[†]

(2) $N = \Sigma N_i$ = constant; the total number of electrons in the system is fixed.

(3) $E_{TOT} = \Sigma E_i N_i$ = constant; the total energy of the system is fixed.

As outlined previously, and subject to the cited constraints, the task at hand is to determine the most likely arrangement of the N electrons in the E_1, E_2, E_3, $\cdots$ energy-level system. The N_i values thereby determined can then be divided by S_i and equated to the value of the Fermi function at E_i—that is, $f(E_i) = N_i$(most probable)$/S_i$.

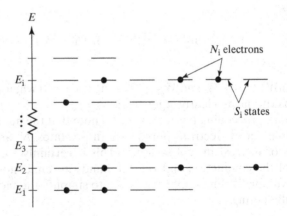

Figure 4.5 Envisioned multi-level energy system of a totally arbitrary nature which contains S_i states and N_i electrons at an energy E_i ($i = 1, 2, 3, \cdots$).

[†] The *Pauli Exclusion Principle* from quantum mechanics dictates that no two electrons in a system can be characterized by the same set of quantum numbers.

Derivation Proper

Consider first the number of different ways (W_i) in which the N_i indistinguishable electrons in the ith level can be placed into the available S_i states. By direct computation, or preferably by reference to a "probability and statistics" textbook, one finds

$$W_i = \frac{S_i!}{(S_i - N_i)! N_i!} \qquad (4.28)$$

If more than one level is considered, the number of different arrangements increases as the product of the individual W_i-values. Since Eq. (4.28) holds for any E_i level, the total number of different ways (W) in which the N electrons can be arranged in the multi-level system is therefore concluded to be

$$W = \prod_i W_i = \prod_i \frac{S_i!}{(S_i - N_i)! N_i!} \qquad (4.29)$$

It should be pointed out that the above W expression is valid for any set of N_i values meeting the $\Sigma N_i = N$ and $\Sigma E_i N_i = E_{TOT}$ restrictions. We seek the set of N_i values for which W is at its maximum. This can be obtained in the usual fashion by setting the total differential of W equal to zero and solving for W_{MAX}. The maximization procedure is drastically simplified, however, if $d(\ln W) = 0$ replaces $dW = 0$ as the maximization criteria. As deduced from Eq. (4.29),

$$\ln W = \sum_i [\ln S_i! - \ln(S_i - N_i)! - \ln N_i!] \qquad (4.30)$$

Since $d(\ln W) = dW/W$ and $W_{MAX} \neq 0$, $d(\ln W) = 0$ when $dW = 0$ and the two maximization criteria are clearly equivalent.

Before proceeding it is important to note that the number of available states, S_i, and the number of electrons populating those states, N_i, are typically quite large for E_i values of interest in real systems. (This is certainly true for the near-band-edge portion of the conduction and valence bands in semiconductors.) We can therefore justify invoking Stirling's approximation to simplify the factorial terms in Eq. (4.30). Specifically noting

$$\ln x! \simeq x \ln x - x \qquad \ldots \text{Stirling's approximation} \qquad (4.31)$$
$$(x \text{ large})$$

one can write

$$\ln W \simeq \sum_i \left[S_i \ln S_i - S_i - (S_i - N_i)\ln (S_i - N_i) + (S_i - N_i) - N_i \ln N_i + N_i \right] \quad (4.32a)$$

$$= \sum_i \left[S_i \ln S_i - (S_i - N_i)\ln (S_i - N_i) - N_i \ln N_i \right] \quad (4.32b)$$

We are finally in a position to perform the actual maximization. Recognizing that $dS_i = 0$ (the S_i are system constants), one obtains

$$d(\ln W) = \sum_i \frac{\partial \ln W}{\partial N_i} \, dN_i \quad (4.33a)$$

$$= \sum_i \left[\ln(S_i - N_i) + 1 - \ln N_i - 1 \right] dN_i \quad (4.33b)$$

$$= \sum_i \ln(S_i/N_i - 1) \, dN_i \quad (4.33c)$$

Setting $d(\ln W) = 0$ then yields

$$\sum_i \ln(S_i/N_i - 1) \, dN_i = 0 \quad (4.34)$$

The solution of Eq. (4.34) for the most probable N_i value set is of course subject to the $\sum N_i = N$ and $\sum E_i N_i = E_{TOT}$ restrictions. These solution constraints can be recast into the equivalent differential form

$$\sum_i dN_i = 0 \quad (4.35a)$$

and

$$\sum_i E_i \, dN_i = 0 \quad (4.35b)$$

To solve Eq. (4.34) subject to the Eq. (4.35) constraints, we employ the method of Lagrange multipliers.[†] This method consists of multiplying each constraint equation by

---

[†]For an excellent discussion of the Lagrange multiplier method (also referred to as the method of undetermined multipliers), see L. P. Smith, *Mathematical Methods for Scientists and Engineers*, Dover Publications, Inc., New York, 1953, pp. 39–42.

an as yet unspecified constant. (Let the undetermined multipliers be $-\alpha$ and $-\beta$, respectively.) The resulting equations are then added to Eq. (4.34), giving

$$\sum_i [\ln (S_i/N_i - 1) - \alpha - \beta E_i] \, dN_i = 0 \tag{4.36}$$

In principle, α and β can always be chosen such that two of the bracketed dN_i coefficients vanish, thereby eliminating two of the dN_i from Eq. (4.36). We assume this to be the case. With two of the dN_i eliminated, all of the remaining dN_i in Eq. (4.36) can be varied independently, and the summation will vanish for all choices of the independent differentials only if

$$\ln(S_i/N_i - 1) - \alpha - \beta E_i = 0 \qquad \dots \text{all } i \tag{4.37}$$

Equation (4.37) is the sought-after relationship for the most probable N_i. Solving Eq. (4.37) for N_i/S_i, we therefore conclude

$$f(E_i) = \frac{N_i}{S_i} = \frac{1}{1 + e^{\alpha + \beta E_i}} \tag{4.38}$$

For closely spaced levels, as encountered in the conduction and valence bands of semiconductors, E_i may be replaced by the continuous variable E and

$$f(E_i) \rightarrow f(E) = \frac{1}{1 + e^{\alpha + \beta E}} \tag{4.39}$$

Concluding Discussion

To complete the derivation of the Fermi function it is necessary to evaluate the solution constants, α and β. This is usually accomplished by performing supplemental theoretical analyses or by comparing the general form of the result with experimental data. Thermodynamic arguments and the analysis of real systems using statistical mechanics or the kinetic theory of gases, for example, lead to the conclusion that

$$\alpha = -\frac{E_F}{kT}$$

and

$$\beta = \frac{1}{kT}$$

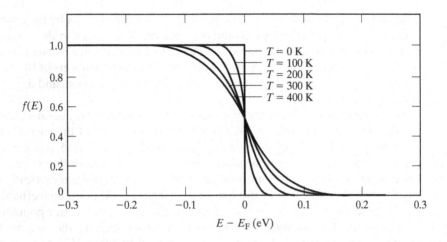

Figure 4.6 Value of the Fermi function versus energy with the system temperature as a parameter.

where E_F is the electrochemical potential or Fermi energy of the electrons in the solid, $k = 8.617 \times 10^{-5}$ eV/K is Boltzmann's constant,[†] and T is the system temperature. Thus we arrive at the final form of the Fermi function

$$f(E) = \frac{1}{1 + e^{(E-E_F)/kT}}$$

(4.40)

A sample plot of the Fermi function versus $E - E_F$ for a select number of temperatures is displayed in Fig. 4.6.

4.3 SUPPLEMENTAL INFORMATION

At this point we could proceed directly to the development of relationships used to calculate the carrier concentrations. Unfortunately, the development and attendant discussions depend to a large extent on an assumed familiarity with what might be classified as carrier and semiconductor modeling. This section, a factual bridge, has been included to supply the necessary supplemental information.

4.3.1 Equilibrium Distribution of Carriers

Having established the energy distribution of available band states and the ratio of filled to total states under equilibrium conditions, one can now easily deduce the distribution of carriers in the conduction and valence bands. The desired distribution is

---

[†]k is widely employed as the symbol of choice for both the wavenumber and Boltzmann's constant. The correct interpretation of the symbol can invariably be determined from the context of an analysis.

obtained by simply multiplying the appropriate density of states by the appropriate oc-cupancy factor: $g_c(E)f(E)$ yields the distribution of electrons in the conduction band and $g_v(E)[1 - f(E)]$ yields the distribution of holes (unfilled states) in the valence band. Sample carrier distributions for different assumed positions of the Fermi energy (along with the associated energy band diagram, Fermi function, and density of states) are pictured in Fig. 4.7.

Two important observations should be made with reference to Fig. 4.7. The first relates to the general form of the carrier distributions. All of the carrier distributions are zero at the band edges, reach a peak close to E_c or E_v, and then decay very rapidly toward zero as one moves upward into the conduction band or downward into the va-lence band. In other words, most of the carriers are grouped energetically in the near vicinity (within a few kT) of the band edges. The second point concerns the effect of the Fermi level positioning on the relative magnitude of the carrier populations. When E_F is positioned in the upper half of the band gap (or higher), the electron population greatly outweighs the hole population. Conversely, when E_F is positioned below midgap, the hole population far outweighs the electron population. Positioning E_F near midgap yields an approximately equal number of electrons and holes. This behavior stems of course from the change in the occupancy factors as a function of E_F.

4.3.2 The Energy Band Diagram

The energy band diagram, introduced in Fig. 4.7 without comment (actually Fig. 4.1(a) is also an energy band diagram), is the workhorse of semiconductor models. Device analyses make extensive use of this diagram. Because we plan to make use of the dia-gram in subsequent discussions, it is useful at this point to review the salient features and interpretations of this exceptional visualization aid.

Basically, referring to Fig. 4.8(a), the energy band diagram is a plot of the allowed electron energy states in a material as a function of *position* along a preselected direc-tion. (This diagram should not be confused with the *E-k* plots of Chapter 3.) In its sim-plest form the diagram contains only two lines: one designating the bottom of the conduction band and the other identifying the top of the valence band. The *x*- and *y*-axis labels shown in Fig. 4.8(a) are routinely omitted in working versions of the dia-gram. It is also implicitly understood that most of the states above E_c are empty, and that most of the states below E_v are filled with electrons.

As pictured in Fig. 4.8(b), filled-in circles, representing electrons, and empty cir-cles, representing holes, are sometimes added to the diagram for conceptual purposes. Arrows drawn adjacent to the carrier representations, as in Fig. 4.8(c), convey an envi-sioned motion of the carriers within the material.

The remaining parts of Fig. 4.8 are all concerned with diagram interpretations of an energy-related nature. Figure 4.8(d) points out that hole energies increase *downward* on the diagram. As noted in Fig. 4.8(e), the kinetic energy (K. E.) of a carrier is equal to the energy displacement between the carrier's position in a band and the band edge; i.e., for electrons K. E. $= E - E_c$, and for holes K. E. $= E_v - E$. This interpretation follows from the fact that the energy supplied to excite an at-rest valence band electron from E_v to E_c is totally used up in the excitation process. The $E = E_c$ electron and $E = E_v$ hole thereby created are therefore at rest. Additional electron or hole carrier energies are

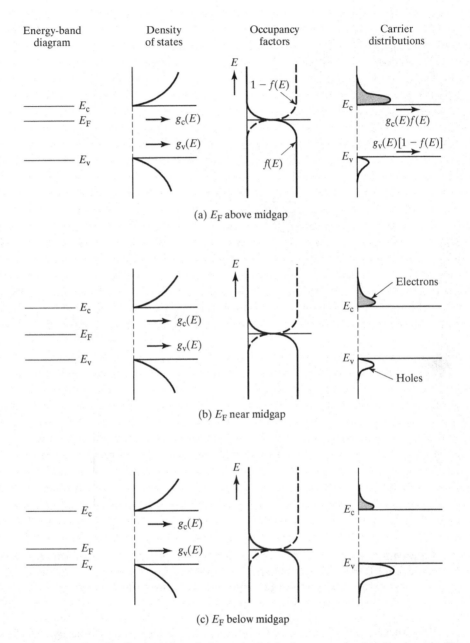

Figure 4.7   Carrier distributions (not drawn to scale) in the respective bands when the Fermi level is positioned (a) above midgap, (b) near midgap, and (c) below midgap. Also shown in each case are coordinated sketches of the energy-band diagram, density of states, and the occupancy factors (the Fermi function and one minus the Fermi function).

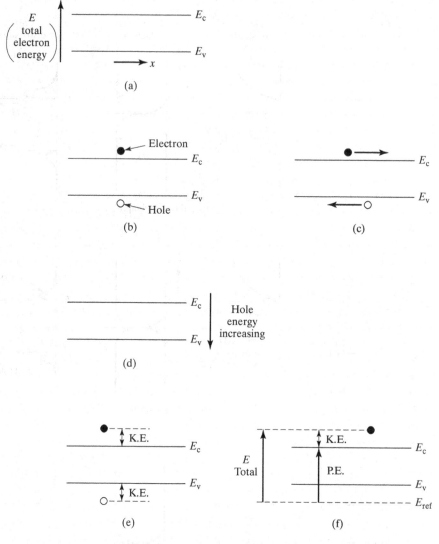

Figure 4.8   The energy band model or diagram. (a) Basic diagram. (b) Representation of carriers. (c) Representation of carrier motion. Diagram interpretation of (d) increasing hole energies, (e) carrier kinetic energies, and (f) the carrier potential energy.

thus logically associated with the energy of motion—kinetic energy. Finally, since the kinetic energy plus the potential energy (P. E.) is equal to the total energy, the electron potential energy as pictured in Fig. 4.8(f) must be equal to $E_c - E_{ref}$. The potential energy (and hence the total energy) is of course arbitrary to within a constant, and the position-independent reference energy, E_{ref}, may be chosen to be any convenient value.

When an electric field ($\mathscr{E}$) exists inside a material, the band energies become a function of position. The resulting variation of E_c and E_v with position on the energy band diagram, exemplified by Fig. 4.9(a), is popularly referred to as "band bending." Combining the diagram band bending and the previously established relationship between the potential energy and E_c, it is possible to deduce, by inspection, the general form of the electrostatic variables inside the material. Specifically, we know from elementary physics that the potential energy of a $- q$ charged particle is related to the electrostatic potential V at a given point by

$$\text{P. E.}(x) = -qV(x) \tag{4.41}$$

Equation (4.41) is valid, of course, provided that only electrostatic forces are acting on the particle. We assume this to be the case. Thus, having previously concluded

$$\text{P. E.} = E_c - E_{\text{ref}} \tag{4.42}$$

we can state

$$\boxed{V = -\frac{1}{q}(E_c - E_{\text{ref}})} \tag{4.43}$$

Moreover, by definition,

$$\mathscr{E} = -\nabla V \tag{4.44}$$

or, in one dimension,

$$\mathscr{E} = -\frac{dV}{dx} \tag{4.45}$$

Differentiating Eq. (4.43) therefore yields

$$\boxed{\mathscr{E} = \frac{1}{q}\frac{dE_c}{dx} = \frac{1}{q}\frac{dE_v}{dx}} \tag{4.46}$$

If the Eq. (4.43) and Eq. (4.46) results are applied to the sample diagram of Fig. 4.9(a), one deduces a V versus x and $\mathscr{E}$ versus x dependence of the form shown in Fig. 4.9(b) and (c), respectively. V is obtained by simply turning the E_c or E_v versus x dependence "upside down"; $\mathscr{E}$ is obtained by merely noting the slope of the band-edge energies as a function of position. Also, since the charge density, ρ, is directly proportional to $d\mathscr{E}/dx$, the procedure can always be extended one step further to obtain ρ versus x. Graphical differentiation of the $\mathscr{E}$ vs. x dependence of Fig. 4.9(c) yields the ρ vs. x dependence of Fig. 4.9(d).

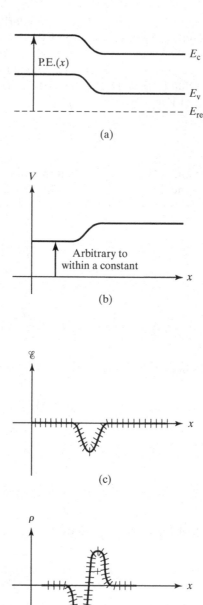

Figure 4.9   Relationship between "band bending" and the electrostatic variables inside a semiconductor. (a) Sample energy band diagram exhibiting band bending. (b) Electrostatic potential, (c) electric field, and (d) charge density versus position deduced from and associated with the part (a) diagram.

As a final point we should note that working versions of the energy band diagram, like those in Fig. 4.10, typically contain two levels in addition to E_c and E_v—namely, the Fermi level (or energy), E_F, and the intrinsic Fermi level, E_i. The intrinsic Fermi level is the energy the Fermi level would assume if there were an equal concentration of electrons and holes in the material (the situation in intrinsic or pure semiconductors). In the discussion on the equilibrium distribution of carriers we found that a near-midgap positioning of E_F gives rise to an approximately equal number of electrons and holes. This should explain why E_i is routinely drawn midway between E_c and E_v on energy band diagrams. With both E_i and E_F on an energy band diagram it is possible to tell at a glance the dominant carrier concentration within a semiconductor, or at a given point in a semiconductor. Figure 4.10(a), for example, models an n-type semiconductor, a semiconductor where electrons are the majority or more populous carrier. Similarly, Fig. 4.10(b) represents a p-type semiconductor, a semiconductor where holes are the majority carrier. In Fig. 4.10(c), on the other hand, the carrier population is depicted as progressively changing from predominantly holes on the left to predominantly electrons on the right. Also note in Fig. 4.10(c) that E_i tracks E_c and E_v—that is, E_i retains the same relative position compared to the band edges even in the presence of band bending.

4.3.3 Donors, Acceptors, Band Gap Centers

A semiconductor sample containing an insignificant amount of impurity atoms is referred to as an *intrinsic semiconductor*. In intrinsic semiconductors the electron and hole carrier populations are always equal. The excitation of a valence band electron into the conduction band simultaneously creates both a carrier electron (henceforth simply referred to as an "electron") and a hole. Since this is the only mechanism available for carrier creation

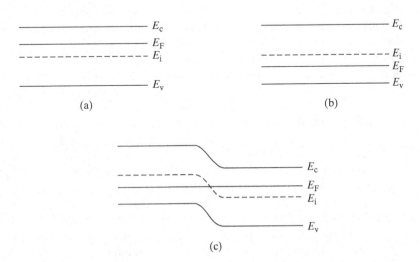

Figure 4.10 Working versions of the energy band diagram containing E_F and E_i. Representations of (a) an n-type semiconductor, (b) a p-type semiconductor, and (c) a pn junction.

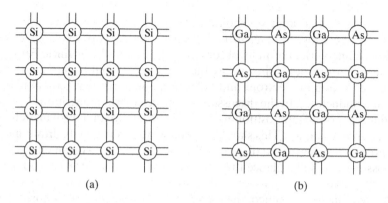

Figure 4.11   The bonding model. Models for (a) a column IV semiconductor exemplified by Si and (b) a III–V compound semiconductor exemplified by GaAs.

in intrinsic materials, the carrier concentrations must be equal. Obviously, the carrier concentrations must be manipulated to achieve the *n*- and *p*-type semiconductors employed in the fabrication of devices. This is accomplished by *doping* the semiconductor—by adding controlled amounts of special impurity atoms known as *donors* and *acceptors*. Donors are added to enhance the electron concentration; acceptors are added to increase the hole concentration.

Donor and acceptor action can best be understood with the aid of the semiconductor *bonding model* shown in Fig. 4.11. In this model the atomic cores (atoms less valence electrons) of the semiconductor are schematically represented by a two-dimensional array of circles; the valence or bonding electrons are represented by lines interconnecting the circles. For a column IV semiconductor such as Si there are four valence electrons per atom and four nearest neighbors (see Fig. 1.5(a)). Each of the valence electrons is equally shared with a nearest neighbor. Thus there are a total of eight lines in the bonding model terminating on each circle. In compound semiconductors such as GaAs, the actual bonding is partly ionic and partly covalent; i.e., the three valence electrons supplied by Ga atoms and the five valence electrons supplied by As atoms are not equally shared between the atoms. The model of Fig. 4.11(b), however, is adequate for most purposes. Note that a bound valence electron in the bonding model corresponds to a valence band electron in the energy band model. Likewise, an electron freed by the breaking of a bond and a missing bond are the bonding model equivalents of a conduction band electron and a valence band hole, respectively.

If one now conceptually substitutes a column V atom, such as phosphorus, for a Si atom (see Fig. 4.12(a)), four of the five phosphorus atom valence electrons will readily be incorporated into the Si bonding structure. The fifth valence electron, however, cannot be incorporated into the bonding scheme, is rather weakly bound to the phosphorus site, and is easily freed from the site by the absorption of thermal energy at room temperature. Effectively, the column V atom "donates" a conduction band electron to the system. The explanation of acceptor action follows a similar line of reasoning. A column III atom such as boron has three valence electrons. This atom cannot complete one of the bonds when substituted for a Si atom in the semiconductor lattice. The column III atom, however, readily accepts an electron from a nearby Si–Si bond (see

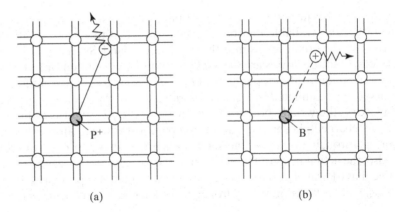

(a)   (b)

Figure 4.12   Visualization of (a) donor and (b) acceptor action using the bonding model. In (a) the column V element P is substituted for a Si atom; in (b) the column III element B is substituted for a Si atom.

Fig. 4.12(b)), thereby completing its own bonding scheme and in the process creating a hole that can wander about the lattice. Note that in both instances there is an increase in only one type of carrier. Charge balance is nevertheless maintained because the fixed donor and acceptor sites themselves become charged (ionized) in providing carriers to the system.

The same general ideas apply to the doping of compound and alloy semiconductors. The column VI element Te when substituted for As yields n-type GaAs; the column II element Zn when substituted for Ga yields p-type GaAs. Elements lying in a column of the periodic table intermediate between the columns containing the semiconductor components can function as either donors or acceptors depending on the lattice site occupied. If both donor- and acceptor-like behavior is observed, the dopant is said to be *amphoteric*. Si in GaAs is a prime example of an amphoteric dopant. Si typically substitutes for Ga, yielding n-type material. Under certain growth conditions, however, Si can be made to substitute for As, yielding p-type material.

In terms of the energy band diagram, donors add allowed electron states in the band gap close to the conduction band edge as pictured in Fig. 4.13(a); acceptors add

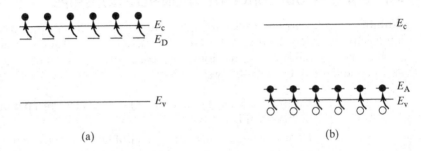

(a)   (b)

Figure 4.13   Energy-band-diagram representation of (a) donors and (b) acceptors. Total ionization of the dopant sites is pictured in the diagrams.

allowed states just above the valence-band edge as shown in Fig. 4.13(b). At room temperature the available thermal energy is sufficient to excite essentially all of the electrons on the donor levels into the conduction band. Similarly, holes are created when valence band electrons are thermally excited into the acceptor levels. The energy E_I required to ionize the dopant sites ($E_I = E_c - E_D$ for donors and $E_A - E_v$ for acceptors) can be estimated by a simple modification of the hydrogen atom analysis. Crudely speaking, the weakly bound fifth electron of the phosphorus atom in Si and the positively charged phosphorus site form a pseudo-hydrogen atom. The electron, however, orbits through a material, the semiconductor, with a non-unity dielectric constant K_S. Also, the actual mass of the electron in the pseudo-atom must be replaced by an effective mass. Parallel arguments can be presented for acceptor sites. Thus, replacing ε_0 by $K_S \varepsilon_0$ and m_0 by m^* in the hydrogen atom analysis, one concludes from Eq. (2.7) that

$$E_I \simeq \frac{m^* q^4}{2(4\pi K_S \varepsilon_0 \hbar)^2} = \frac{13.6}{K_S^2} \frac{m^*}{m_0} \; (\text{eV}) \qquad (4.47)$$

For most semiconductors, $K_S \sim 10$ and $m^*/m_0 \lesssim 1$, giving $E_I \lesssim 0.1$ eV. This estimate is in agreement with experimental results and justifies positioning the donor and acceptor levels close to the band edges.

In addition to the relatively shallow-level donors and acceptors, the substitution of other impurity atoms into a host semiconductor can give rise to deep-level states—allowed states in the band gap more than a few tenths of an eV from either band edge. These are commonly referred to as traps or recombination-generation centers. Such centers are usually unintentional impurities, are not necessarily ionized at room temperature, and typically occur in concentrations far below the dopant concentration. The deep-level centers can be donor-like (positively charged when empty and neutral when filled with an electron), acceptor-like (neutral when empty and negatively charged when filled with an electron), or may exhibit multiple charge states with associated multiple levels in the forbidden gap. Figure 4.14 summarizes the observed band gap levels (donor, acceptor, and deep level) introduced by the most commonly encountered impurities in Ge, Si, and GaAs.

4.4   EQUILIBRIUM CONCENTRATION RELATIONSHIPS

With the required groundwork completed, we can now develop the computational relationships routinely used to determine the carrier concentrations inside a semiconductor under equilibrium conditions. The concentration symbols employed in the development are defined as follows:

n ...   Electron concentration; total number of electrons per cm^3 in the conduction band.

p ...   Hole concentration; total number of holes per cm^3 in the valence band.

n_i ...   Intrinsic carrier concentration; electron and hole concentration in intrinsic material.

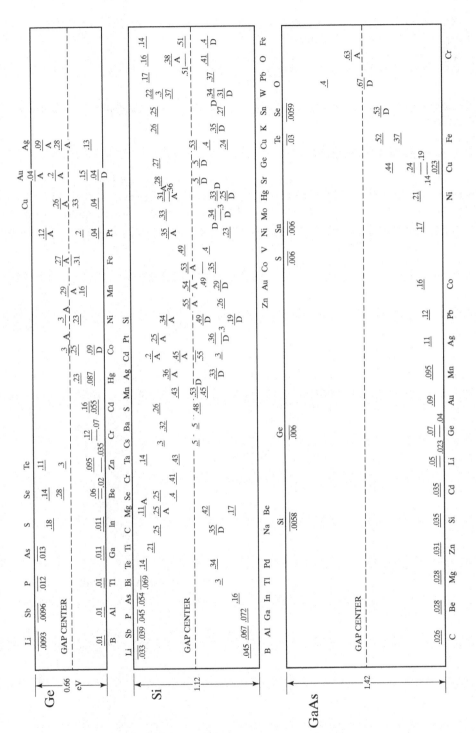

Figure 4.14 Measured ionization energies for the most commonly encountered impurities in Ge, Si, and GaAs. The levels above midgap are referenced to E_c and are donor-like or multiply charged donors, unless marked with an A which identifies an acceptor level. The levels below midgap are referenced to E_v and are acceptor-like or multiply charged acceptors, unless marked with a D for donor level. (From Sze.[3] Reprinted with permission.)

N_D^+ ...   Number of ionized donors per cm^3
N_D ...   Total donor concentration
N_A^- ...   Number of ionized acceptors per cm^3
N_A ...   Total acceptor concentration

Please note that, in general, each of the above quantities except n_i can be a function of position inside the semiconductor.

4.4.1   Formulas for *n* and *p*

The number of electrons/cm^3 and holes/cm^3 with energies between E and $E + dE$ has been established to be $g_c(E)f(E)dE$ and $g_v(E)[1 - f(E)]dE$, respectively. The total carrier concentration in a band is therefore obtained by simply integrating the appropriate distribution function over the energy band—that is,

$$n = \int_{E_c}^{E_{top}} g_c(E)f(E)dE \tag{4.48a}$$

$$p = \int_{E_{bottom}}^{E_v} g_v(E)[1 - f(E)]dE \tag{4.48b}$$

Substituting the density of states and Fermi function expressions into Eqs. (4.48), noting that little error is introduced by letting $E_{bottom} \to -\infty$ and $E_{top} \to \infty$, and rearranging the result into a convenient form, one obtains

$$n = N_C \mathcal{F}_{1/2}(\eta_c) \tag{4.49a}$$

$$p = N_V \mathcal{F}_{1/2}(\eta_v) \tag{4.49b}$$

where

$$N_C = 2\left(\frac{2\pi m_n^* kT}{h^2}\right)^{3/2} \quad \ldots \text{effective density of conduction band states} \tag{4.50a}$$

$$N_V = 2\left(\frac{2\pi m_p^* kT}{h^2}\right)^{3/2} \quad \ldots \text{effective density of valence band states} \tag{4.50b}$$

$$\mathcal{F}_{1/2}(\eta) = \frac{2}{\sqrt{\pi}} F_{1/2}(\eta) \tag{4.51}$$

$$F_{1/2}(\eta) = \int_0^\infty \frac{\xi^{1/2} d\xi}{1 + e^{\xi - \eta}} \quad \ldots \text{Fermi-Dirac integral of order 1/2} \tag{4.52}$$

and

$$\eta_c = (E_F - E_c)/kT \tag{4.53a}$$

$$\eta_v = (E_v - E_F)/kT \tag{4.53b}$$

The Eq. (4.49) concentration relationships are valid for any conceivable positioning of the Fermi level. N_C and N_V, the effective density of states, are of course readily computed for a given material and temperature: the 300 K values of these constants for Ge, Si, and GaAs are listed in Table 4.2A. [At 300 K, $N_{C,V} = (2.510 \times 10^{19}/cm^3)(m^*_{n,p}/m_0)^{3/2}.] \mathscr{F}_{1/2}(\eta)$, on the other hand, is obtained from literature tabulations, through direct computation, or by the use of analytical approximations. Selected properties of the $\mathscr{F}_j(\eta)$ family of modified Fermi–Dirac integrals is presented in Table 4.2B. Of the several analytical approximations for $\mathscr{F}_{1/2}(\eta$ known)

Table 4.2 Concentration Parameters and Functions

A. Effective density of states at 300 K

Semiconductor	$N_C\,(cm^{-3})$	$N_V\,(cm^{-3})$
Ge	1.03×10^{19}	5.35×10^{18}
Si	3.23×10^{19}	1.83×10^{19}
GaAs	4.21×10^{17}	9.52×10^{18}

B. Selected properties of the $\mathscr{F}_j(\eta)$ functions [4]

$$\mathscr{F}_j(\eta) \equiv \frac{1}{\Gamma(j+1)} \int_0^\infty \frac{\xi^j d\xi}{1 + e^{\xi - \eta}}$$

$$\mathscr{F}_j(\eta) \rightarrow e^\eta \quad \text{as} \quad \eta \rightarrow -\infty$$

$$\frac{d}{d\eta}\mathscr{F}_j(\eta) = \mathscr{F}_{j-1}(\eta)$$

$$\mathscr{F}_{1/2}(\eta) \simeq [e^{-\eta} + \xi(\eta)]^{-1}$$

where $\xi(\eta) = 3\sqrt{\pi/2}\,[(\eta + 2.13) + (|\eta - 2.13|^{2.4} + 9.6)^{5/12}]^{-3/2}$

with a maximum error of $\sim\pm 0.5\%$

$$\eta \simeq \frac{\ln u}{1 - u^2} + \frac{(3\sqrt{\pi}u/4)^{2/3}}{1 + [0.24 + 1.08(3\sqrt{\pi}u/4)^{2/3}]^{-2}}$$

where $u \equiv \mathscr{F}_{1/2}(\eta)$

with a maximum error of $\sim \pm 0.5\%$

and $\eta(\mathscr{F}_{1/2}$ known) suggested in the device literature,[4,5] the entries in this table provide a reasonable combination of accuracy and convenience. The asymtotic approach of $\mathscr{F}_{1/2}(\eta)$ to $\exp(\eta)$ as η increases negatively (entry #2 in Table 4.2B) is examined in detail in Fig. 4.15. For additional information about the $\mathscr{F}_{1/2}(\eta)$ function and a summary of available $\mathscr{F}_{1/2}(\eta)$ versus η tabulations, the reader is referred to the excellent review paper by Blakemore.[4]

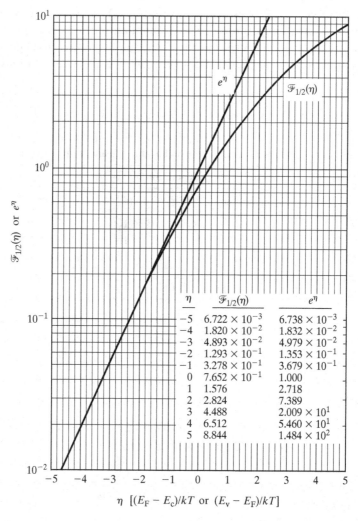

η	$\mathscr{F}_{1/2}(\eta)$	e^{η}
−5	6.722×10^{-3}	6.738×10^{-3}
−4	1.820×10^{-2}	1.832×10^{-2}
−3	4.893×10^{-2}	4.979×10^{-2}
−2	1.293×10^{-1}	1.353×10^{-1}
−1	3.278×10^{-1}	3.679×10^{-1}
0	7.652×10^{-1}	1.000
1	1.576	2.718
2	2.824	7.389
3	4.488	2.009×10^{1}
4	6.512	5.460×10^{1}
5	8.844	1.484×10^{2}

η $[(E_F - E_c)/kT$ or $(E_v - E_F)/kT]$

Figure 4.15   Comparison of the modified Fermi–Dirac integral and e^{η} for η near zero (corresponding to Fermi level positionings near the band edges).

As is evident from Fig. 4.15, $\mathscr{F}_{1/2}(\eta)$ is closely approximated by $\exp(\eta)$ when $\eta \leq -3$. Utilizing this approximation, one obtains

$$n = N_C e^{(E_F - E_c)/kT} \qquad \ldots E_c - E_F \geq 3kT \ (\eta_c \leq -3) \qquad (4.54a)$$

$$p = N_V e^{(E_v - E_F)/kT} \qquad \ldots E_F - E_v \geq 3kT \ (\eta_v \leq -3) \qquad (4.54b)$$

The inequalities adjacent to Eqs. (4.54) are simultaneously satisfied if the Fermi level lies in the band gap more than $3kT$ from either band edge. For the cited positioning of the Fermi level (also see Fig. 4.16), the semiconductor is said to be *nondegenerate* and Eqs. (4.54) are referred to as nondegenerate relationships. Conversely, if the Fermi level is within $3kT$ of either band edge or lies inside a band, the semiconductor is said to be *degenerate*. It should be noted that a nondegenerate positioning of the Fermi level makes $f(E) \simeq \exp[-(E - E_F)/kT]$ for all conduction band energies and $1 - f(E) \simeq \exp[(E - E_F)/kT]$ for all valence band energies. The simplified form of the occupancy factors is a Maxwell–Boltzmann type function that also describes, for example, the energy distribution of molecules in a high-temperature, low-density gas. When substituted into Eqs. (4.48), the simplified occupancy factors lead directly to the nondegenerate relationships.

Although in closed form, the Eq. (4.54) relationships find limited usage in device analyses. More often than not one employs an equivalent set of relationships involving a reduced number of system parameters and energy levels. Since the nondegenerate relationships are obviously valid for an intrinsic semiconductor where $n = p = n_i$ and $E_F = E_i$, one can write

$$n_i = N_C e^{(E_i - E_c)/kT} \qquad (4.55a)$$

$$n_i = N_V e^{(E_v - E_i)/kT} \qquad (4.55b)$$

or, solving for the effective density of states,

$$N_C = n_i e^{(E_c - E_i)/kT} \qquad (4.56a)$$

$$N_V = n_i e^{(E_i - E_v)/kT} \qquad (4.56b)$$

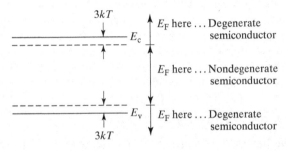

Figure 4.16 Definition of degenerate/nondegenerate semiconductors.

Eliminating N_C and N_V in the original nondegenerate relationships using Eqs. (4.56) then yields

$$n = n_i e^{(E_F - E_i)/kT} \tag{4.57a}$$

$$p = n_i e^{(E_i - E_F)/kT} \tag{4.57b}$$

Like Eqs. (4.54), the more convenient (4.57) expressions are valid for any semiconductor in equilibrium whose doping is such as to give rise to a nondegenerate positioning of the Fermi level.

4.4.2 n_i and the np product

As can be inferred from its appearance in Eqs. (4.57), the intrinsic carrier concentration figures prominently in the quantitative calculation of the carrier concentrations. It is therefore reasonable to interject considerations involving this important material constant.

If the corresponding sides of Eqs. (4.54a) and (4.54b) are multiplied together, one obtains

$$np = N_C N_V e^{(E_v - E_c)/kT} = N_C N_V e^{-E_G/kT} \tag{4.58}$$

A similar multiplication of the corresponding sides of Eqs. (4.57a) and (4.57b) yields

$$np = n_i^2 \tag{4.59}$$

Thus, equating the right-hand sides of the preceding np expressions and solving for n_i, we conclude

$$n_i = \sqrt{N_C N_V}\, e^{-E_G/2kT} \tag{4.60}$$

Although appearing trivial, the np production relationship [Eq. (4.59)] often proves to be extremely useful in practical computations. Given one of the carrier concentrations, the remaining concentration is readily determined using Eq. (4.59). Eq. (4.60), on the other hand, expresses n_i as a function of known quantities and may be used to compute n_i at a specified temperature or as a function of temperature. The best available plots of n_i as a function of temperature for Ge, Si, and GaAs are displayed in Fig. 4.17.

4.4.3   Charge Neutrality Relationship

Thus far we have yet to relate the carrier concentrations to the dopants present in a semiconductor. In general the connection between the two quantities can be quite

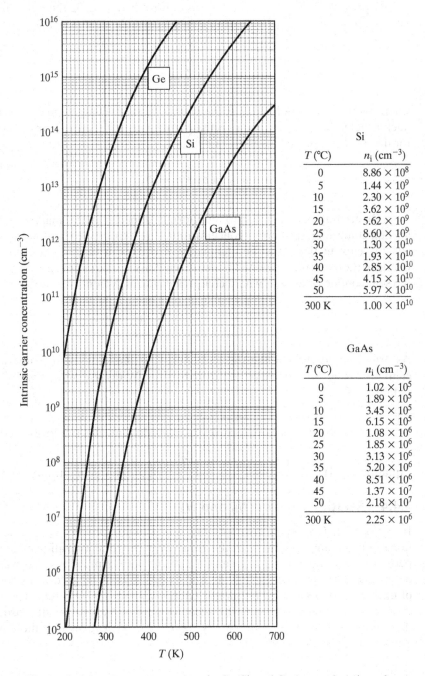

Si	
T (°C)	n_i (cm^{-3})
0	8.86×10^8
5	1.44×10^9
10	2.30×10^9
15	3.62×10^9
20	5.62×10^9
25	8.60×10^9
30	1.30×10^{10}
35	1.93×10^{10}
40	2.85×10^{10}
45	4.15×10^{10}
50	5.97×10^{10}
300 K	1.00×10^{10}

GaAs	
T (°C)	n_i (cm^{-3})
0	1.02×10^5
5	1.89×10^5
10	3.45×10^5
15	6.15×10^5
20	1.08×10^6
25	1.85×10^6
30	3.13×10^6
35	5.20×10^6
40	8.51×10^6
45	1.37×10^7
50	2.18×10^7
300 K	2.25×10^6

Figure 4.17 Intrinsic carrier concentrations in Ge, Si, and GaAs as a function of temperature.

involved. A very simple relationship can be established, however, if equilibrium conditions prevail and the semiconductor is uniformly doped. Seeking the special-case relationship, let us begin by writing down Poisson's equation from electromagnetic theory:

$$\nabla \cdot \mathscr{E} = \rho/K_S \varepsilon_0 \qquad (4.61)$$

where, as previously defined, ρ is the charge density (charge/cm^3), K_S is the semiconductor dielectric constant, and ε_0 is the permittivity of free space. Inside a semiconductor the local charge density is given by[†]

$$\rho = q(p - n + N_D^+ - N_A^-) \qquad (4.62)$$

Restricting our attention to a uniformly doped material maintained under equilibrium conditions, the carrier concentrations are expected to be uniform with position and the current flow must be identically zero. This in turn implies that $\mathscr{E} = 0$ at all points in the semiconductor and hence, from Poisson's equation, $\rho = 0$. In other words, the semiconductor is everywhere charge-neutral, and, from Eq. (4.62),

$$\boxed{p - n + N_D^+ - N_A^- = 0} \qquad (4.63)$$

As previously discussed, the thermal energy available at room temperature is sufficient to ionize almost all of the dopant sites. With $N_D^+ = N_D$ and $N_A^- = N_A$, one obtains

$$\boxed{p - n + N_D - N_A = 0} \qquad \text{...dopant sites totally ionized} \qquad (4.64)$$

Equation (4.64) is the normally quoted form of the charge neutrality relationship.

4.4.4   Relationships for N_D^+ and N_A^-

In performing $T = 300$ K computations it is common practice to assume total ionization of the dopant sites. To check the accuracy of the total ionization assumption and to perform low-temperature computations, we require expressions that will allow us to determine the degree of dopant-site ionization. Since N_D^+/N_D corresponds to the ratio of empty to total states at the donor energy E_D, and N_A^-/N_A represents the ratio of electron filled to total states at the acceptor energy E_A, one might reasonably expect that $N_D^+/N_D = 1 - f(E_D)$ and $N_A^-/N_A = f(E_A)$. As it turns out, the correct dopant ionization expressions are

---

[†]In writing down Eq. (4.62) we implicitly assumed that the deep-level trap concentrations (N_{T1}, N_{T2}, etc.) inside the semiconductor were negligible compared to the net doping concentration. If this is not the case, terms of the form $+qN_T^+$ or $-qN_T^-$ must be added to Eq. (4.62).

$$\frac{N_D^+}{N_D} = \frac{1}{1 + g_D e^{(E_F - E_D)/kT}} \qquad \cdots g_D = 2 \text{ (standard value)} \qquad (4.65a)$$

$$\frac{N_A^-}{N_A} = \frac{1}{1 + g_A e^{(E_A - E_F)/kT}} \qquad \cdots g_A = 4 \text{ (standard value)} \qquad (4.65b)$$

where g_D and g_A are the donor- and acceptor-site *degeneracy factors*, respectively. (It should be noted that the term "degeneracy" in the present context implies equivalent multiplicity and bears no relationship to the degenerate or nondegenerate positioning of the Fermi level.)

Although very similar to the expressions expected from a straightforward extrapolation of earlier results, the dopant ionization relationships contain additional factors multiplying the exponential term in the Fermi function. This arises because the statistics of filling band gap levels differs slightly from the statistics of filling energy band levels. In the Fermi function derivation it was assumed that each allowed state could accommodate one and only one electron characterized by a specific set of quantum numbers. A donor site, for example, can still accommodate only one electron. However, the donor-site electron can have either a spin-up or a spin-down. This fact increases the ways of arranging electrons on the donor-level sites relative to energy levels in the bands.

To obtain the Eq. (4.65a) result, a special E_D level must be added to the Fermi function derivation. Each of the states at the energy E_D is taken to be g_D-fold degenerate and the W expression in the Fermi function derivation, Eq. (4.29), is replaced by

$$W' = W W_D \qquad (4.66)$$

$$W_D = \frac{g_D^{N_D - N_D^+} N_D!}{(N_D - N_D^+)! N_D^+!} \qquad (4.67)$$

W_D is the number of different ways in which electrons can be arranged on the donor sites. Completion of the derivation following the procedure given in Section 4.2 then yields Eq. (4.65a). The Eq. (4.65b) acceptor relationship can be obtained in a similar manner.

A few comments are in order concerning the values and handling of the degeneracy factors. Most authors set $g_D = 2$ to account for the spin degeneracy at the donor sites. It is likewise standard practice to argue that the acceptor sites must additionally reflect the two-fold (heavy- and light-hole) degeneracy of the valence band. Thus the acceptor degeneracy factor is usually taken to be $g_A = 2 \times 2 = 4$. However, more detailed theoretical analyses suggest that the situation may be more complex. For example, there is the possibility of excited bound states (like in the hydrogen atom) giving rise to increased degeneracy. One can in fact find a range of degeneracy factors, even fractional degeneracy factors, in the device literature. This is particularly true for deep level donor-like and acceptor-like traps whose ionization expressions have a form analogous to Eqs. (4.65). [For deep level donor-like centers, $N_D^+ \rightarrow N_T^+$, $g_D \rightarrow g_T$, and

$E_D \rightarrow E_T$ in Eq. (4.65a); for deep level acceptor-like centers, $N_A^- \rightarrow N_T^-$, $g_A \rightarrow g_T$, and $E_A \rightarrow E_T$ in Eq. (4.65b).]

The accurate experimental determination of the degeneracy factors also tends to be rather elusive and is often intertwined with the determination of a center's energy level. Note that one can always write

$$\frac{N_T^+}{N_T} = \frac{1}{1 + g_T e^{(E_F - E_T)/kT}} = \frac{1}{1 + e^{(E_F - E_T')/kT}} \tag{4.68}$$

$$E_T' \equiv E_T - kT \ln g_T \tag{4.69}$$

which would be appropriate for a deep level donor-like trap. More often than not it is E_T' (or analogously E_D', E_A') that is determined directly from experimental data. Some authors simply assume a g-value—often $g = 1$ for deep-level traps—and quote an E-value based on the g assumption. The g assumption is lost, of course, in preparing plots like Fig. 4.14. Foregoing a search of the original literature, it is advisable to interpret the Fig. 4.14 values to be $E_D (g_D = 2)$, $E_A (g_A = 4)$, and $E_T' (g_T = 1)$ for the donors, acceptors, and deep-level centers, respectively.

4.5   CONCENTRATION AND E_F CALCULATIONS

4.5.1   General Information

With the fundamental relationships established, it is simply a matter of combining and manipulating these relationships to determine the carrier concentrations and to locate the position of the Fermi level inside a given semiconductor sample. Practically speaking, once one has determined n, p, or E_F, the values of the two remaining variables are readily determined from the fundamental relationships. In this section we perform the necessary manipulations to obtain the computational expressions routinely employed in concentration and E_F calculations. A limited number of sample calculations are also presented. It is assumed throughout that *equilibrium conditions prevail* and that the material under analysis is either an *undoped or uniformly doped nondegenerate semiconductor.*

To facilitate simplifications during the course of the calculations it is useful to have a general idea as to the expected temperature dependence of the majority carrier concentration inside a doped semiconductor. For illustrative purposes let us consider an N_D-doped ($N_D \neq 0$, $N_A = 0$) semiconductor where $N_D \gg n_i$ at room temperature. Figure 4.18(a) typifies the expected n versus T dependence in the specified material. At temperatures approaching zero Kelvin there is insufficient thermal energy in the material to excite electrons from the valence band into the conduction band or even to ionize the donor sites. Thus, as visualized on the left-hand side of Fig. 4.18(b), $n \rightarrow 0$ as $T \rightarrow 0$ K. Raising the temperature slightly above zero Kelvin causes some of the donor sites to ionize. However, the energy available is still too low to excite a significant number of electrons across the band gap. Hence, the number of observed electrons at low temperatures (formally named the "freeze-out" temperature region) is equal to the number of ionized donors, $n \simeq N_D^+$. Further increasing the semiconductor temperature eventually causes almost complete ionization of the donor sites. Since by

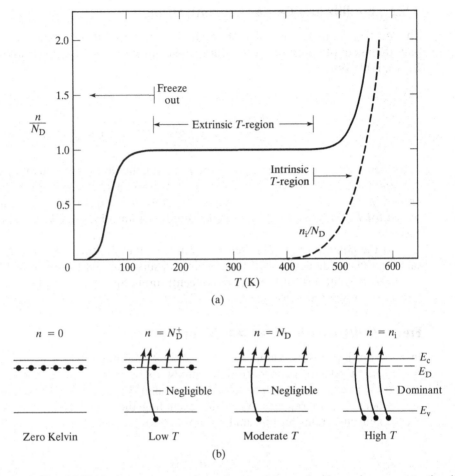

Figure 4.18 (a) Typical temperature dependence of the majority-carrier concentration in a doped semiconductor. The plot was constructed assuming a phosphorus-doped $N_D = 10^{15}/cm^3$ Si sample. n_i/N_D versus T (dashed line) has been included for comparison purposes. (b) Qualitative explanation of the concentration-versus-temperature dependence displayed in part (a).

assumption $N_D \gg n_i$ in the given material at room temperature, the contribution to the electron population from band-to-band excitation clearly remains negligible at least up to room temperature. It therefore follows that $n \simeq N_D$ = constant over the temperature range where $N_D^+ \simeq N_D$ and $N_D \gg n_i$. This is referred to as the extrinsic temperature region and is the normal operating region for solid-state devices. Finally, if the temperature is increased above room temperature, the electrons excited across the band gap eventually approach, then exceed, and, as pictured on the right-hand side of Fig. 4.18(b), ultimately swamp the fixed number of electrons derived from the donor sites. The temperature range where $n \simeq n_i$ is referred to as the intrinsic temperature region.

4.5.2   Equilibrium Carrier Concentrations

The starting point for all calculations is the charge neutrality relationship. Making only the assumption of nondegeneracy, one can substitute Eqs. (4.54) and (4.65) into Eq. (4.63) to obtain

$$N_V e^{(E_v - E_F)/kT} - N_C e^{(E_F - E_c)/kT} + \frac{N_D}{1 + g_D e^{(E_F - E_D)/kT}} - \frac{N_A}{1 + g_A e^{(E_A - E_F)/kT}} = 0 \quad (4.70)$$

In a given problem, the temperature, the material, the dominant dopant center or centers, and the dopant concentrations are all taken to be known quantities; the only unknown in Eq. (4.70) is E_F. Very generally, therefore, Eq. (4.70) can be numerically solved for E_F and the value of E_F substituted back into Eq. (4.54), thereby yielding n and p.

   If the computation is limited to a range of temperatures, it is often possible to simplify the charge neutrality relationship and subsequently to obtain highly accurate closed-form solutions for the carrier concentrations. Specific examples of practical interest are considered below.

Freeze-Out/Extrinsic T ($N_D \gg N_A$ or $N_A \gg N_D$)

In a donor-doped semiconductor ($N_D \gg N_A$) maintained at temperatures where $N_D \gg n_i$, the electron concentration will always be much greater than the hole concentration. Likewise, N_D^+ will be much greater than N_A^- except in the extreme $T \to 0$ K limit where N_D^+ approaches N_A. Thus, excluding the $T \to 0$ K limit if $N_A \neq 0$, the charge neutrality relationship can be simplified to

$$n = N_D^+ \quad (4.71)$$

This result is of course in agreement with the previous qualitative discussion. Using Eqs. (4.65a) and (4.54a) we can also write

$$N_D^+ = \frac{N_D}{1 + g_D e^{(E_F - E_D)/kT}} = \frac{N_D}{1 + g_D(n/N_C)e^{(E_c - E_D)/kT}} \quad (4.72a)$$

$$= \frac{N_D}{1 + (n/N_\zeta)} \quad (4.72b)$$

where

$$N_\zeta \equiv (N_C/g_D)e^{-(E_c - E_D)/kT} \quad \text{(a computable constant at a given } T) \quad (4.73)$$

Eliminating N_D^+ in Eq. (4.71) using Eq. (4.72b) and solving for n, one obtains

$$n = \frac{N_D}{1 + (n/N_\zeta)} \tag{4.74}$$

$$n^2 + N_\zeta n - N_\zeta N_D = 0 \tag{4.75}$$

and

$$n = -\frac{N_\zeta}{2} + \left[\left(\frac{N_\zeta}{2}\right)^2 + N_\zeta N_D\right]^{1/2} \quad \begin{array}{l}(+ \text{ root chosen} \\ \text{because } n \geq 0)\end{array} \tag{4.76a}$$

or

$$n = \frac{N_\zeta}{2}\left[\left(1 + \frac{4N_D}{N_\zeta}\right)^{1/2} - 1\right] \tag{4.76b}$$

An analogous result can be obtained for acceptor-doped material.

Upon examining Eq. (4.76b), note that N_ζ will typically be much greater than N_D in the extrinsic temperature region and $n \to N_D$. For example, taking the semiconductor to be $N_D = 10^{15}/\text{cm}^3$ phosphorus-doped Si and $T = 300$ K, $E_c - E_D = 0.045$ eV, $g_D = 2$, $N_C = 3.226 \times 10^{19}/\text{cm}^3$, $N_\zeta = 2.829 \times 10^{18}/\text{cm}^3$, and from Eq. (4.76b) one computes $n = 0.9996 N_D$. Since $n = N_D^+$, this result also tells us the phosphorus donor sites in $N_D = 10^{15}/\text{cm}^3$ Si are 99.96% ionized at room temperature and supports the usual total-ionization approximation for room-temperature operation. By way of comparison, the donor sites in the same semiconductor are only 73.4% ionized at liquid-nitrogen temperatures. ($N_C = 3.57 \times 10^{18}/\text{cm}^3$ and $N_\zeta = 2.02 \times 10^{15}/\text{cm}^3$ at 77 K if one employs the 4 K m_n^* for Si listed in Table 4.1.) It should also be mentioned that the entire low-temperature portion of the Fig. 4.18(a) plot was constructed using Eq. (4.76b).

Extrinsic/Intrinsic T

For a semiconductor maintained at a temperature where the vast majority of dopant sites are ionized, the charge neutrality relationship simplifies to

$$p - n + N_D - N_A = 0 \tag{4.77}$$
$$\text{(Same as 4.64)}$$

In a nondegenerate semiconductor, however, $np = n_i^2$. Thus we have

$$n_i^2/n - n + N_D - N_A = 0 \tag{4.78a}$$

or

$$n^2 - (N_D - N_A)n - n_i^2 = 0 \qquad (4.78b)$$

Solving the quadratic equation for n then yields

$$n = \frac{N_D - N_A}{2} + \left[\left(\frac{N_D - N_A}{2} \right)^2 + n_i^2 \right]^{1/2} \quad \begin{array}{l} (+ \text{ root chosen} \\ \text{because } n \geq 0) \end{array} \qquad (4.79a)$$

and

$$p = n_i^2/n = \frac{N_A - N_D}{2} + \left[\left(\frac{N_A - N_D}{2} \right)^2 + n_i^2 \right]^{1/2} \qquad (4.79b)$$

When a semiconductor is maintained in the extrinsic temperature region, $N_D \gg n_i$ in a donor-doped ($N_D \gg N_A$) semiconductor and $N_A \gg n_i$ in an acceptor-doped ($N_A \gg N_D$) semiconductor. Thus for extrinsic temperature operation, which normally includes room temperature, Eqs. (4.79) reduce to

$$\boxed{\begin{array}{l} n \simeq N_D \\ p \simeq n_i^2/N_D \end{array}} \qquad \begin{array}{l} \text{donor-doped, extrinsic-}T \\ (N_D \gg N_A, N_D \gg n_i) \end{array} \qquad \begin{array}{l} (4.80a) \\ (4.80b) \end{array}$$

$$\begin{array}{l} p \simeq N_A \\ n \simeq n_i^2/N_A \end{array} \qquad \begin{array}{l} \text{acceptor-doped, extrinsic-}T \\ (N_A \gg N_D, N_A \gg n_i) \end{array} \qquad \begin{array}{l} (4.80c) \\ (4.80d) \end{array}$$

Likewise, in the intrinsic temperature region, where $n_i \gg |N_D - N_A|$, Eqs. (4.79) simplify to

$$\boxed{\begin{array}{l} n \simeq n_i \\ p \simeq n_i \end{array}} \qquad \begin{array}{l} \text{intrinsic-}T \\ (n_i \gg |N_D - N_A|) \end{array} \qquad \begin{array}{l} (4.81a) \\ (4.81b) \end{array}$$

The results here are, of course, in total agreement with the earlier qualitative discussion. In the extrinsic temperature range, the majority carrier concentration is simply equal to the dominant doping concentration, and the minority carrier concentration equals n_i^2 divided by the dominant doping concentration; $N_D = 10^{15}/\text{cm}^3$ doped Si at $T = 300$ K would have $n \simeq 10^{15}/\text{cm}^3$ and $p \simeq 10^5/\text{cm}^3$. Moreover, regardless of the doping, all semiconductors ultimately become intrinsic at sufficiently elevated temperatures. Note that the complete expressions, Eqs. (4.79), need be employed only for temperatures where $n_i \sim |N_D - N_A|$.

Finally, to increase the resistivity, donors or acceptors are sometimes added to make $N_D - N_A \simeq 0$; in other materials, such as GaAs, N_A may be comparable to N_D in the as-grown crystal. When N_A and N_D are comparable and non-zero, the material is said to be *compensated*, with the effects of the dopants tending to negate each other. If this be the case, both N_D and N_A must be retained in the carrier concentration expressions.

4.5.3   Determination of E_F

The position of the Fermi level is often determined as an adjunct to carrier concentration calculations. For one, the Fermi level positioning is sometimes needed to confirm the validity of the nondegenerate assumption. Knowledge of the Fermi level positioning is also desired in drawing energy band diagrams. The precise nondegenerate positioning of E_F can always be computed, of course, from Eq. (4.70). (If it turns out that the semiconductor is degenerate, the E_F value thereby determined will extend further into the degenerate zone than the true E_F value.) Like the carrier concentrations, however, highly accurate closed-form solutions for E_F are possible in most practical cases of interest. Specific examples are considered below.

Exact Position of E_i

Given an intrinsic ($N_A = 0$, $N_D = 0$) semiconductor, one can write

$$n = p \tag{4.82}$$

or, making use of Eqs. (4.54),

$$N_C e^{(E_F - E_c)/kT} = N_V e^{(E_v - E_F)/kT} \tag{4.83}$$

Solving for $E_F = E_i$ yields

$$E_i = \frac{E_c + E_v}{2} + \frac{kT}{2} \ln (N_V/N_C) \tag{4.84}$$

and since

$$\frac{N_V}{N_C} = \left(\frac{m_p^*}{m_n^*} \right)^{3/2} \tag{4.85}$$

we conclude that

$$E_i = \frac{E_c + E_v}{2} + \frac{3}{4} kT \ln (m_p^*/m_n^*) \tag{4.86}$$

Using the effective masses listed in Table 4.1, one finds E_i in Si is positioned 0.0073 eV below midgap and E_i in GaAs is positioned 0.0403 eV above midgap at 300 K. Thus, at 300 K, the energy displacement of E_i from midgap is 0.65% and 2.8% of the band gap energy for Si and GaAs, respectively.

Freeze-Out/Extrinsic T ($N_D \gg N_A$ or $N_A \gg N_D$)

In a donor-doped nondegenerate semiconductor where $N_D \gg n_i$, we know

$$n = N_C e^{(E_F - E_c)/kT} = (N_\zeta/2)[(1 + 4N_D/N_\zeta)^{1/2} - 1] \qquad (4.87)$$

giving

$$E_F = E_c + kT\ln\{(N_\zeta/2N_C)[(1 + 4N_D/N_\zeta)^{1/2} - 1]\} \qquad (4.88)$$

Equation (4.88) is particularly useful for low-temperature calculations. As can be verified using Eq. (4.88), E_F rises toward the conduction band edge when T is decreased, approaching a limiting value midway between E_c and E_D as $T \to 0$ K. (If $N_A \neq 0$, E_F approaches E_D.) Analogously, in acceptor-doped semiconductors, E_F approaches $(E_A + E_v)/2$ if $N_D = 0$ and E_A if $N_D \neq 0$ in the $T \to 0$ K limit.

Extrinsic/Intrinsic T

When the semiconductor temperature is maintained in the extrinsic/intrinsic temperature regions it is more convenient to work with the Eq. (4.57) n and p expressions involving n_i. Solving Eqs. (4.57) for $E_F - E_i$, one obtains

$$E_F - E_i = kT\ln(n/n_i) = -kT\ln(p/n_i) \qquad (4.89)$$

Depending on the simplifications inherent in a particular problem, the appropriate extrinsic/intrinsic carrier concentration solution [Eqs. (4.79), (4.80), or (4.81)] can then be substituted into Eq. (4.89) to determine the positioning of E_F. Note that $E_F \to E_i$ in the intrinsic temperature region, as must be the case. Also, for typical device operating temperatures and semiconductor doping conditions,

$$\boxed{\begin{array}{ll} E_F - E_i = kT\ln(N_D/n_i) & \dots N_D \gg N_A, \; N_D \gg n_i \qquad (4.90a) \\[2ex] E_i - E_F = kT\ln(N_A/n_i) & \dots N_A \gg N_D, \; N_A \gg n_i \qquad (4.90b) \end{array}}$$

A plot of E_F versus T for select Si doping concentrations constructed using the relationships developed in this subsection is displayed in Fig. 4.19.

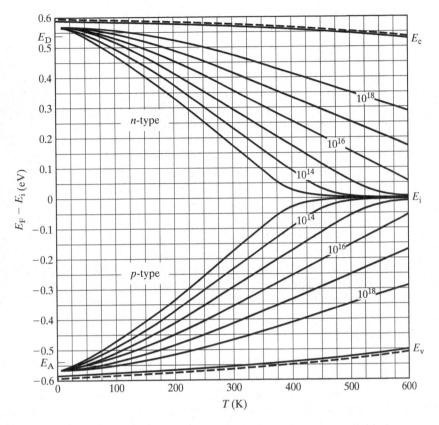

Figure 4.19   Fermi level positioning in Si as a function of temperature for select doping concentrations. The doping concentrations chosen span the range from the minimum readily attainable in Si to the degenerate limit. The dopant ionization energies were taken to be 0.045 eV, $g_D = 2$, $g_A = 4$, and the Si was assumed to contain only one type of dopant.

4.5.4   Degenerate Semiconductor Considerations

Although we do not intend to examine degenerate semiconductor computations in any detail, a few comments are nevertheless in order. At first glance it would appear that degeneracy could be taken into account by simply replacing the nondegenerate carrier relationships in the foregoing analyses with the more general Eq. (4.49) relationships. Unfortunately, the required modifications are much more extensive. Degeneracy is typically caused by heavy doping, with N_D or N_A in excess of approximately $10^{18}/cm^3$ in Si at room temperature. When the semiconductor is heavily doped, there arises a number of *many-body effects* associated with the large majority carrier concentration. For one, a large electron or hole concentration acts to screen the charge on the respective ionized impurity sites. This leads to a reduction in the impurity ionization energy ($E_c - E_D$ or $E_A - E_v$). At dopings only slightly greater than the degenerate limit, the ionization energy goes to zero, and the impurity level moves into the nearby energy band. Additional many-body effects arise from the majority-carrier/majority-carrier and majority-carrier/minority-carrier interactions. Specifically, at large majority carrier concentrations these interactions lead to a narrowing of the band gap, with the conduction band edge decreasing in energy and the valence band edge increasing in energy.

Coincident with the occurrence of many-body effects there is a second set of effects associated with the statistically random nature of the impurity ion distribution. The impurity ion distribution is never perfectly uniform but varies somewhat from point to point within the semiconductor. At large dopant concentrations, this causes significant fluctuations in the local electrostatic potential, which in turn gives rise to a spatial variation in the local density of states distribution. When the local density of states distribution is averaged over the entire lattice, the resulting macroscopic density of states used in defining the macroscopic properties of the semiconductor exhibits *band tails*, allowed energy states "tailing" into the band gap from the conduction and valence bands.

Even in the presence of many body and band tailing effects, one can still compute the majority carrier concentration for operation in the extrinsic temperature region employing $n \cong N_D$ in n-type material and $p \cong N_A$ in p-type material. However, one cannot compute the intrinsic carrier concentration, compute the minority carrier concentration, or accurately determine the position of the Fermi level with the relationships developed herein. For detailed information about heavy doping effects and the associated computational modifications, the reader is referred to the semiconductor literature.[6]

REFERENCES

[1] H. D. Barber, "Effective Mass and Intrinsic Concentration in Silicon," Solid-State Electronics, *10*, 1039 (1967).

[2] J. S. Blakemore, "Semiconducting and Other Major Properties of Gallium Arsenide," J. Appl. Phys., *53*, R123 (Oct., 1982).

[3] S. M. Sze, *Physics of Semiconductor Devices*, 2nd edition, John Wiley & Sons. Inc., New York, 1981.

[4] J. S. Blakemore, "Approximations for Fermi-Dirac Integrals, Especially the Function $\mathscr{F}_{1/2}(\eta)$ Used to Describe Electron Density in a Semiconductor," Solid-State Electronics, *25*, 1067 (1982).

[5] S. A. Wong, S. P. McAlister, and Z.-M. Li, "A Comparison of Some Approximations for the Fermi-Dirac Integral of Order 1/2," Solid-State Electronics, *37*, 61 (1994).

[6] R. J. Van Overstraeten and R. P. Mertens, "Heavy Doping Effects in Silicon," Solid-State Electronics, *30*, 1077 (1987).

PROBLEMS

4.1 A particle of mass m and fixed energy E is confined to a *two-dimensional* box. The x and y side-lengths of the box are a and b, respectively. Also $U(x, y) =$ constant everywhere inside the box. Assuming the side-lengths of the box are much larger than atomic dimensions, derive an expression for the density of states (g_{2D}) for the given particle in a two-dimensional box. Record all steps in obtaining your answer.

4.2 The E-k relationship characterizing electrons confined to a two-dimensional surface layer is of the form

$$E - E_c = \frac{\hbar^2 k_x^2}{2m_1} + \frac{\hbar^2 k_y^2}{2m_2} \quad \cdots m_1 \neq m_2$$

Assuming the side-lengths of the surface layer are much larger than atomic dimensions, derive an expression for the density of states (g_{2D}) for the electrons in the two-dimensional surface layer. Record all steps in obtaining your answer.

4.3 A particle of mass m and fixed energy E is confined to a special three-dimensional box where the z *side-length is on the order of atomic dimensions*. The other two box dimensions are much larger than atomic dimensions. As in the text density-of-states derivation the x, y, and z side-lengths are a, b, c, respectively (see Fig. P4.3); $U(x, y, z) =$ constant everywhere inside the box.

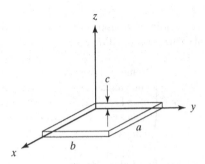

Figure P4.3

(a) Does the small size of the z-dimension have any effect on the overall wavefunction solution embodied in Eqs. (4.6) through (4.9)?

(b) What effect will the small size of the z-dimension have on the k-space representation of Schrödinger equation solutions (Fig. 4.3a)? Make a rough sketch of the revised k-space representation of Schrödinger equation solutions. DO NOT include trivial (k_x, k_y or $k_z = 0$) solutions on your revised plot.

(c) The density of states in this problem will exhibit discontinuities at energies $E_n = (\hbar^2/2m)(n\pi/c)^2$, $n = 1, 2, 3, \cdots$.
  (i) Using the part (b) sketch, explain why discontinuities in the density of states occur at the stated E_n values.
  (ii) Establish expressions for the density of states in the energy ranges $0 \leq E \leq E_1$, $E_1 \leq E \leq E_2$, $E_n \leq E \leq E_{n+1}$.
      HINT: The segmented density of states solution is related to the g_{2D} of Problem 4.1.

(d) On the same set of coordinates plot both $g(E)$ versus E from part (c) and the standard large-dimension result [Eq. (4.17)]. Carefully compare the two results.

[COMMENT: Electrons or holes in an inversion layer near the surface of a semiconductor are effectively confined to a reduced-dimension box similar to the one considered in this problem. The electrons or holes so confined are sometimes referred to as a "two-dimensional gas." (The inversion layer problem is somewhat more complicated, however, in that $U(x, y, z) \neq$ constant.)]

4.4 The conduction band minima in GaP occur right at the first Brillouin zone boundary along the $< 100 >$ directions in k-space. Taking the constant energy surfaces to be ellipsoidal with $m_\ell^*/m_0 = 1.12$ and $m_t^*/m_0 = 0.22$, determine the density of states effective mass for electrons in GaP.

4.5 The valence band of InSb is a bit unusual in that the heavy-hole subband exhibits maxima along $\langle 111 \rangle$ directions at a k-value slightly removed from $k = 0$. If the heavy-hole maxima are described by parabolic energy surfaces where m_ℓ^* and m_t^* are the longitudinal and transverse effective masses, respectively, and if $m_{\ell h}^*$ is the effective mass for the light holes in a spherical subband centered at $k = 0$, obtain an expression for the density of states effective mass characterizing the holes in InSb. Your answer should be expressed in terms of m_ℓ^*, m_t^*, and $m_{\ell h}^*$.

4.6 In Si, where $m_{hh}^*/m_0 = 0.537$ and $m_{\ell h}^*/m_0 = 0.153$, what fraction of the holes are heavy holes? (For simplicity, assume that the quoted 4 K effective masses can be employed at any temperature.)

4.7 In calculating the thermionic emission current flowing through a Schottky barrier diode one needs an analytical expression for d^3n, where

$$d^3n \equiv \begin{pmatrix} \text{Number of conduction band electrons per} \\ \text{unit volume with a } v_x \text{ velocity between} \\ v_x \text{ and } v_x + dv_x, \text{ a } v_y \text{ velocity between} \\ v_y \text{ and } v_y + dv_y, \text{ and a } v_z \text{ velocity} \\ \text{between } v_z \text{ and } v_z + dv_z. \end{pmatrix}$$

Assuming the conduction band electrons in the semiconductor under analysis can be characterized by an isotropic effective mass m_e^*, and noting

$$\hbar k_x = m_e^* v_x; \qquad \hbar k_y = m_e^* v_y; \qquad \hbar k_z = m_e^* v_z$$

develop the required expression for d^3n. As a starting point you may use any result established in the text. Be sure to clearly explain what you are doing.

4.8 Making use of Eq. (4.67), and following the procedure outlined in Section 4.2, derive Eq. (4.65a).

4.9 The carrier distributions or number of carriers as a function of energy in the conduction and valence bands were noted to peak at an energy very close to the band edges. (See the carrier distribution plots in Fig. 4.7.) Taking the semiconductor to be nondegenerate, determine the precise energy relative to the band edges at which the carrier distributions peak.

4.10 (a) Establish a general expression (involving integrals) for the average kinetic energy, $\langle K.E. \rangle$, of the conduction band electrons in a semiconductor.
     (b) Taking the semiconductor to be nondegenerate, simplify your general $\langle K.E. \rangle$ expression to obtain a closed-form result.

4.11 Six different silicon samples maintained at 300 K are characterized by the energy band diagrams in Fig. P4.11. Answer the questions that follow after choosing a specific diagram for analysis. Possibly repeat using other energy band diagrams.
     (a) Sketch the electrostatic potential (V) inside the semiconductor as a function of x.
     (b) Sketch the electric field ($\mathscr{E}$) inside the semiconductor as a function of x.

(c) The carrier pictured on the diagram moves back and forth between $x = 0$ and $x = L$ without changing its total energy. Sketch the kinetic energy and potential energy of the carrier as a function of position inside the semiconductor. Let E_F be the energy reference level.

(d) Roughly sketch n and p versus x.

(e) Is the semiconductor degenerate at any point? If so, where?

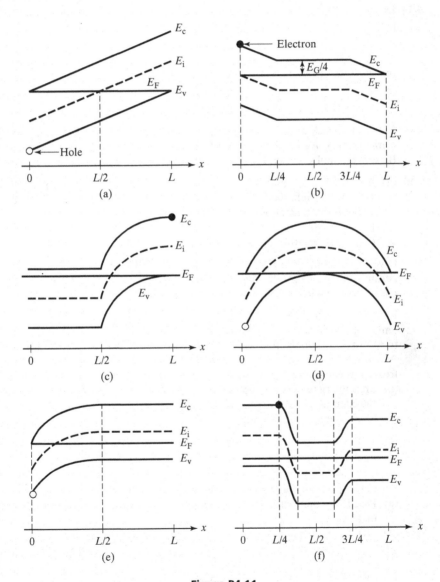

Figure P4.11

4.12 Because m_n^* and thus N_C are relatively small in GaAs, the donor doping at which the material becomes degenerate is also relatively small. Determine the N_D doping at the non-degenerate/degenerate transition point in GaAs at 300 K.

4.13 Determine the degenerate doping limit for donors in Si as a function of temperature at 100 K increments between $T = 200$ K and $T = 600$ K. Employ

$$m_n^*/m_0 = 1.028 + (6.11 \times 10^{-4})\,T - (3.09 \times 10^{-7})T^2$$

4.14 Determine the equilibrium electron and hole concentrations inside a uniformly doped sample of Si under the following conditions:
(a) $T = 300$ K, $N_A \ll N_D$, $N_D = 10^{14}/cm^3$.
(b) $T = 300$ K, $N_A = 10^{15}/cm^3$, $N_D \ll N_A$.
(c) $T = 300$ K, $N_A = 9 \times 10^{15}/cm^3$, $N_D = 10^{16}/cm^3$.
(d) $T = 470$ K, $N_A = 0$, $N_D = 10^{14}/cm^3$.
(e) $T = 645$ K, $N_A = 0$, $N_D = 10^{14}/cm^3$.

4.15 (a to e) For each of the conditions specified in Problem 4.14, determine the position of E_i, compute $E_F - E_i$, and draw a carefully dimensioned energy band diagram for the Si sample. Be sure to employ the correct E_G values at the elevated temperatures in parts (d) and (e).

4.16 (a) A set of five Si samples with decade values of donor doping ranging from $10^{13}/cm^3$ to $10^{17}/cm^3$ are maintained at $T = 545$ K. In all cases, $N_D \gg N_A$. What are the electron and hole concentrations inside the five Si samples?
(b) Determine the position of E_i, compute $E_F - E_i$, and draw a carefully dimensioned energy band diagram for each of the Si samples.

4.17 Given a Si sample phosphorus-doped with $N_D = 10^{15}/cm^3 \gg N_A$, calculate n/N_D for temperatures $T = 25$ K, 50 K, 75 K, and 100 K. Assume $g_D = 2$ and $E_c - E_D = 0.045$ eV. Compare your calculated results with Fig. 4.18(a).

4.18 In a material where $N_D > N_A$ but $N_A \neq 0$, the Fermi level E_F will lie above E_i for any system temperature in the freeze-out and extrinsic ranges—i.e., the material will always appear to be n-type. Suppose that to be specific we consider a nondegenerate Si sample where $N_D > N_A$ but $N_A \neq 0$. Further assume that the system temperature is restricted to the freeze-out and extrinsic ranges.
(a) Present arguments leading to the conclusion that $N_A^- \simeq N_A$ in the Si sample at any T in the cited temperature ranges.
(b) Show that one can also write

$$n \simeq N_D^+ - N_A$$

in the given n-type ($n \gg p$) sample.
(c) Paralleling the text $n = N_D^+$ development, derive an expression for N_D^+ in the $N_A \neq 0$ material.
(d) (i) From the part (b) expression, what is the value of N_D^+ if $T \rightarrow 0$ K?
    (ii) According to your part (c) result, what is N_D^+ if $T \rightarrow 0$ K? Record your reasoning. (Are your answers here consistent?)
    (iii) What is the limiting position of E_F when $T \rightarrow 0$ K? Explain.

4.19 It is standard practice to employ break points in a semilog plot of n versus $1000/T$ to determine the boundaries of the extrinsic temperature region. Suppose instead we define the low temperature limit (T_{min}) of the extrinsic temperature region to be that temperature where $n = 0.9N_D$. Similarly, suppose we define the high temperature limit (T_{max}) to be that temperature where $n = 1.1N_D$. Based on the cited definitions, assuming the semiconductor to be non-degenerately N_D-doped with $N_A = 0$, and freely employing results and/or plots contained in the text:

(a) Indicate how you would proceed in determining T_{min} for a given donor doping.

(b) Determine T_{max} for $N_D = 10^{14}/cm^3$, $10^{15}/cm^3$, and $10^{16}/cm^3$ doped Si.

(c) Determine T_{max} for GaAs dopings of $N_D = 10^{14}/cm^3$, $10^{15}/cm^3$, and $10^{16}/cm^3$.

(d) Since solid-state devices are normally operated in the extrinsic temperature region, what do you conclude about the use of Si and GaAs devices at elevated temperatures?

4.20 In InSb at 300 K, $E_G = 0.18$ eV (the smallest band gap of all semiconductor compounds), $m_n^*/m_0 = 0.0116$, $m_p^*/m_0 = 0.40$, and $n_i = 1.6 \times 10^{16}/cm^3$.

(a) Would you expect the intrinsic Fermi energy (E_i) in InSb to lie closer to E_c or E_v? Present a *qualitative* argument that supports your answer—the text relationship for E_i is NOT to be used.

(b) Assuming nondegenerate statistics, determine the positioning of E_i in the InSb band gap at 300 K.

(c) Draw a dimensioned energy band diagram showing the positioning of E_i determined in part (b). (Numerical values for relevant energy differences are noted on a "dimensioned" diagram.) Do you see anything wrong with the part (b) result? Explain.

(d) If something is wrong with the part (b) result, determine the correct positioning of E_i in the InSb band gap.

(e) Given an InSb sample doped with $10^{14}/cm^3$ donors, what is the approximate positioning of E_F in the sample at 300 K? Please note how you deduced your answer.

CHAPTER 5

Recombination–Generation Processes

When a semiconductor is perturbed from the equilibrium state there is typically an attendant modification in the carrier numbers inside the semiconductor. Recombination–generation (R–G) is nature's order-restoring mechanism, the means whereby the carrier excess or deficit inside the semiconductor is stabilized (if the perturbation is maintained) or eliminated (if the perturbation is removed). Since nonequilibrium conditions prevail during device operation, recombination–generation more often than not plays a major role in shaping the characteristics exhibited by a device. In this chapter we first provide basic R–G information and include a survey of recombination–generation processes. The majority of the chapter is devoted to a detailed examination of the often-dominant R–G center process; both bulk and surface recombination–generation are analyzed. The chapter concludes with a brief presentation of practical R–G facts and supplemental information.

5.1 INTRODUCTION

5.1.1 Survey of R–G Processes

In semiconductor work the terms *Recombination* and *Generation* are defined as follows:

> Recombination: A process whereby electrons and holes (carriers)
> are annihilated or destroyed.
> Generation: A process whereby electrons and holes are created.

These definitions are clearly of a very general nature and actually encompass a number of function-related processes. Herein we survey the more common R–G processes, using the energy band diagram as the prime visualization aid. Because of its particular relevance in optical applications, special note is made of how energy is added to or removed from the system during each of the R–G events.

Recombination Processes

(1) *Band-to-Band Recombination.* Band-to-band recombination, also referred to as direct thermal recombination, is conceptually the simplest of all recombination processes. As pictured in Fig. 5.1(a), it merely involves the direct annihilation of a conduction band electron and a valence band hole, the electron falling from an allowed conduction band state into a vacant valence band state. This process is typically *radiative*, with the excess energy released during the process going into the production of a photon (light).

(2) *R–G Center Recombination.* In Subsection 4.3.3 it was noted that certain impurity atoms can introduce allowed energy levels (E_T) into the midgap region of a semiconductor. Crystal defects, particularly defects "decorated" with impurity atoms, can also give rise to deep-level states. As shown in Fig. 5.1(b), the R–G centers thereby created act as intermediaries in the envisioned recombination process. First one type of carrier and then the other type of carrier is attracted to the R–G center. The capture of an electron and a hole at the same site leads to the annihilation of the electron-hole pair. Alternatively, the process may be described in terms of the state-to-state transitions of a single carrier: a carrier is first captured at the R–G site and then makes an annihilating transition to the opposite carrier band. R–G center recombination, also called indirect thermal recombination,[†] is characteristically non-radiative. Thermal energy (heat) is released during the process, or equivalently, lattice vibrations (phonons) are produced.

(3) *Recombination via Shallow Levels.* Like R–G centers, donor and acceptor sites can also function as intermediaries in the recombination process (see Fig. 5.1(c)). If an electron is captured at a donor site, however, it has a high probability at room temperature of being re-emitted into the conduction band before completing the remaining step(s) of the recombination process. A similar statement can be made for holes captured at acceptor sites. For this reason donor and acceptor sites may be likened to extremely inefficient R–G centers, and the probability of recombination occurring via shallow levels is usually quite low at room temperature. It should be noted, nevertheless, that the largest energy step in shallow-level recombination is typically radiative and that the probability of observing shallow-level processes increases with decreasing system temperature.

(4) *Recombination Involving Excitons.* Normally, electrons and holes may be viewed as individual particles that respond independently to applied forces. However, it is possible for an electron and a hole to become bound together into a hydrogen-atom-like arrangement which moves as a unit in response to applied forces. This coupled electron-hole pair is called an *exciton*. It is also possible for one of the exciton components to be trapped at a shallow-level site; the resulting configuration is called a *bound exciton*. Since a certain amount of energy goes into the

---

[†]In the older device literature, R–G centers and recombination–generation via R–G centers are sometimes referred to as SRH (Shockley, Read, Hall) centers and SRH recombination–generation, respectively. W. Shockley and W. T. Read, Jr[1]., and independently R. N. Hall[2,3], were the first to model and investigate this process.

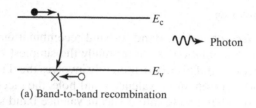

(a) Band-to-band recombination

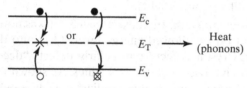

(b) R–G center recombination

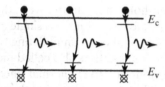

(c) Recombination via shallow levels

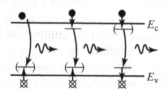

(d) Recombination involving excitons

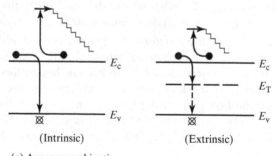

(Intrinsic)   (Extrinsic)

(e) Auger recombination

Figure 5.1 Energy-band visualizations of recombination processes.

formation of an exciton, the difference between the electron and hole energies of the coupled pair can be less than the band gap energy. The formation of an exciton is therefore viewed as introducing a temporary level into the band gap slightly above the valence band edge or slightly below the conduction band edge. In Fig. 5.1(d) these levels are enclosed by parentheses. As pictured in Fig. 5.1(d), recombination of the exciton components can give rise to subband-gap radiation. Recombination involving excitons is a very important mechanism at low temperatures and is the major light-producing mechanism in Light Emitting Diodes (LEDs) containing shallow-level isoelectronic centers.

(5) *Auger Recombination.* The final recombination processes to be considered are the non-radiative Auger (pronounced Oh-jay) type processes. In an Auger process (see Fig. 5.1(e)), band-to-band recombination or trapping at a band gap center occurs simultaneously with the collision between two like carriers. The energy released by the recombination or trapping subprocess is transferred during the collision to the surviving carrier. Subsequently, this highly energetic carrier "thermalizes"— loses energy in small steps through collisions with the semiconductor lattice. The "staircases" in Fig. 5.1(e) represent the envisioned stepwise loss of energy. Because the number of carrier-carrier collisions increases with increased carrier concentration, Auger recombination likewise increases with carrier concentration, becoming very important at high carrier concentrations. Auger recombination must be considered, for example, in treating degenerately doped regions of a device structure and in the detailed operational modeling of concentrator-type solar cells, junction lasers, and LEDs.

Generation Processes

Any of the foregoing recombination processes can be reversed to generate carriers. Band-to-band generation, for example, is pictured in Fig. 5.2(a). Note that either thermal energy or light can provide the energy required for the band-to-band transition. If thermal energy is absorbed the process is alternatively referred to as direct thermal generation; if externally introduced light is absorbed the process is called photogeneration. The thermally assisted generation of carriers with R–G centers acting as intermediaries is envisioned in Fig. 5.2(b). Figure 5.2(c) pictures the photoemission of carriers from band gap centers, typically a rather improbable process. Finally, impact ionization, the inverse of Auger recombination, is shown in Fig. 5.2(d). In this process an electron-hole pair is produced as a result of the energy released when a highly energetic carrier collides with the crystal lattice. The generation of carriers through impact ionization routinely occurs in the high $\mathscr{E}$-field regions of devices and is responsible for the avalanche breakdown in *pn* junctions.

5.1.2 Momentum Considerations

All of the various recombination–generation processes we have cited occur at all times in all semiconductors—they even occur under equilibrium conditions. The critical issue is not whether the processes are occurring, but rather the rates at which the various processes are occurring. Typically, one need be concerned only with the dominant process, the process proceeding at the fastest rate. We have already pointed out that a number of the processes are only important under special conditions or are highly

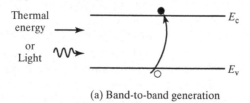

(a) Band-to-band generation

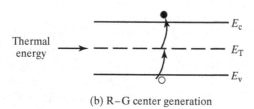

(b) R–G center generation

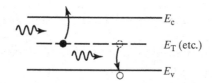

(c) Photoemission from band gap centers

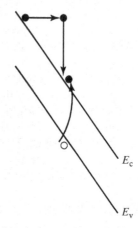

(d) Carrier generation via impact ionization

Figure 5.2   Energy-band visualizations of generation processes.

improbable (occur at a much slower rate than competing processes). Thus, based on our survey, one might expect either band-to-band or R–G center recombination–generation to be dominant in low-field regions of a nondegenerately doped semiconductor maintained at room temperature. From the energy band explanation alone one might even speculate that band-to-band recombination–generation would dominate under the cited "standard" conditions. Visualization of R–G processes using the energy band diagram, however, can be misleading. The E-x plot examines only changes in energy, whereas crystal momentum in addition to energy is conserved in any R–G process. Changes in the crystal momentum must also be examined and, as it turns out, momentum-conservation requirements play an important role in setting the process rate.

Crystal-momentum-related aspects of R–G processes are conveniently discussed with the aid of E-k plots. In Subsection 3.3.2 we noted that semiconductors can be divided into two basic groups depending on the general form of the E-k plot. Direct semiconductors such as GaAs are characterized by E-k plots where the conduction band minimum and the valence band maximum both occur at $k = 0$. In indirect semiconductors such as Si and Ge, the conduction band minimum and the valence band maximum occur at different values of k. The two general plot forms are sketched in Fig. 5.3. To employ these plots in visualizing an R–G process, one also needs to know the nature of transitions associated with the absorption or emission of photons and phonons. Photons, being massless entities, carry very little momentum, and a photon-assisted transition is essentially vertical on an E-k plot. (For GaAs the $\Gamma \rightarrow X$ k-width of the E-k diagram is $2\pi/a$, where a is the GaAs lattice constant; $a = 5.65$ Å at room temperature. By way of comparison, $k_{\text{photon}} = 2\pi/\lambda$. If $E_{\text{photon}} = E_G = 1.42$ eV at room temperature, $\lambda = 0.87\,\mu\text{m}$. Clearly, $\lambda \gg a$ and $k_{\text{photon}} \ll 2\pi/a$.) Conversely, the thermal energy associated with lattice vibrations (phonons) is in the 10–50 meV range, whereas the effective phonon mass and associated momentum are comparatively large. Thus on an E-k plot a phonon-assisted transition is essentially horizontal.

Let us now re-examine the band-to-band recombination process. In a direct semiconductor where the k-values of electrons and holes are all bunched near $k = 0$, little change in momentum is required for the recombination process to proceed. The conservation of both energy and crystal momentum is readily met simply by the emission

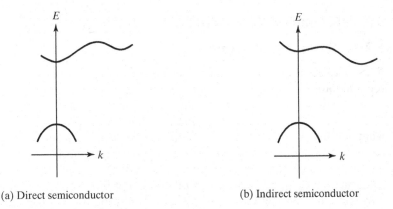

(a) Direct semiconductor   (b) Indirect semiconductor

Figure 5.3   General forms of E-k plots for direct and indirect semiconductors.

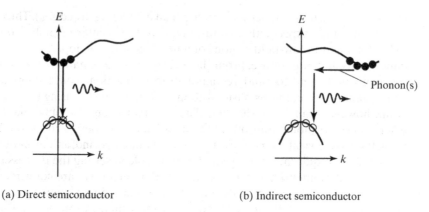

(a) Direct semiconductor   (b) Indirect semiconductor

Figure 5.4 *E-k* plot visualizations of recombination in direct and indirect semiconductors.

of a photon (see Fig. 5.4(a)). In an indirect semiconductor, on the other hand, there is a large change in crystal momentum associated with the recombination process. The emission of a photon will conserve energy but cannot simultaneously conserve momentum. Thus for band-to-band recombination to proceed in an indirect semiconductor a phonon must be emitted (or absorbed) coincident with the emission of a photon (see Fig. 5.4(b)).

The rather involved nature of the band-to-band process in indirect semiconductors understandably leads to a diminished recombination rate. Band-to-band recombination is in fact totally negligible compared to R–G center recombination in indirect semiconductors. Although band-to-band recombination proceeds at a much faster rate in direct semiconductors, the R–G center process can never be neglected and even dominates in many instances. Because of its central and often dominant role in the recombination–generation process, we herein concentrate on the R–G center mechanism, devoting the next two sections to its detailed analysis. The analytical procedures developed, it should be noted, are directly applicable to other R–G processes.

5.2 RECOMBINATION–GENERATION STATISTICS

5.2.1 Definition of Terms

R–G statistics is just the technical name given to the mathematical characterization of recombination–generation processes. Since all R–G processes act to change the carrier concentrations as a function of time, "mathematical characterization" simply means obtaining expressions for $\partial n/\partial t$ and $\partial p/\partial t$ due to the process under consideration. In what follows we concentrate on obtaining relationships for $\partial n/\partial t$ and $\partial p/\partial t$ due to recombination–generation via a single-level center. That is, the semiconductor under analysis is assumed to contain only one type of R–G center which introduces allowed states at an energy E_T into the central portion of the band gap. Actual semiconductors may contain a number of deep-level centers, but the process is typically dominated by a single center.

For the purposes of analysis let us define:

$\left.\dfrac{\partial n}{\partial t}\right|_{R-G}$ ... Time rate of change in the electron concentration due to *both* R–G center recombination and R–G center generation.

$\left.\dfrac{\partial p}{\partial t}\right|_{R-G}$ ... Time rate of change in the hole concentration due to R–G center recombination–generation.

n_T ... Number of R–G centers per cm^3 that are filled with electrons (equivalent to the previously employed N_T^- if the centers are acceptor-like or $N_T - N_T^+$ if the centers are donor-like).

p_T ... Number of empty R–G centers per cm^3.

N_T ... Total number of R–G centers per cm^3, $N_T = n_T + p_T$.

It should be emphasized that $\partial n/\partial t|_{R-G}$ and $\partial p/\partial t|_{R-G}$ are net rates, taking into account the effects of both recombination and generation. $\partial n/\partial t|_{R-G}$ will be negative if there is a net loss of electrons (R > G) or positive if there is a net gain of electrons (G > R). The designation "$|_{R-G}$" indicates that the carrier concentrations are changing "due to recombination–generation via R–G centers." The "due to" designation is necessary because, in general, the time rate of change of the carrier concentrations can be affected by a number of processes, including non-R–G processes.

5.2.2 Generalized Rate Relationships

Consider the possible R–G center to energy band transitions shown in Fig. 5.5. The possible transitions, four in all, are (a) electron capture at an R–G center, (b) electron emission from an R–G center, (c) hole capture at an R–G center, and (d) hole emission from an R–G center. The latter two transitions may alternatively be thought of as

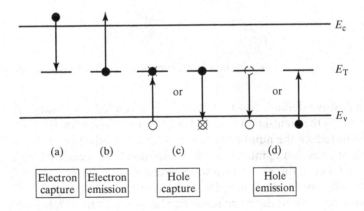

Figure 5.5 Possible electronic transitions between a single-level R–G center and the energy bands.

(c) an electron trapped at an R–G center falling into a vacant valence band state and (d) a valence band electron being excited to the R–G level. The former descriptions of valence band interactions (c) and (d) in terms of holes, however, are preferred and more convenient for our purposes. Since only transitions (a) and (b) affect the electron concentration, and only transitions (c) and (d) affect the hole concentration, one can obviously write

$$
\left.\frac{\partial n}{\partial t}\right|_{R-G} = \left.\frac{\partial n}{\partial t}\right|_{(a)} + \left.\frac{\partial n}{\partial t}\right|_{(b)}
\tag{5.1a}
$$

$$
\left.\frac{\partial p}{\partial t}\right|_{R-G} = \left.\frac{\partial p}{\partial t}\right|_{(c)} + \left.\frac{\partial p}{\partial t}\right|_{(d)}
\tag{5.1b}
$$

As is evident from Eqs. (5.1), fundamental process rates (a) to (d) must be expressed in terms of basic system variables to characterize mathematically the overall recombination–generation rates.

Examining fundamental process (a) in greater detail, we note that electron capture depends critically on the existence of electrons to be captured and the availability of empty R–G centers. If either the electron concentration or the empty-center concentration goes to zero, $\partial n/\partial t|_{(a)} \to 0$. Moreover, if the nonzero concentration of either electrons or empty R–G centers is conceptually doubled, the probability of capturing an electron doubles and $\partial n/\partial t|_{(a)}$ likewise doubles. In other words, process (a) is expected to proceed at a rate *directly* proportional to both n and p_T. Taking c_n to be the constant of proportionality, we can therefore write

$$
\left.\frac{\partial n}{\partial t}\right|_{(a)} = -c_n p_T n
\tag{5.2}
$$

where

$$c_n \ldots \text{the electron capture coefficient(cm}^3/\text{sec)}$$

is a positive-definite constant. Since c_n is taken to be a positive quantity, a minus sign is added to the right-hand side of Eq. (5.2) to account for the fact that electron capture acts to reduce the number of electrons in the conduction band. Note from a consideration of units that c_n must have the dimensions of 1/(concentration-time) or cm^3/sec.

Later in this section we will present an alternative (and perhaps more satisfying) derivation of Eq. (5.2) based on collision theory. At this point it is nevertheless worthwhile to note that the same result can be obtained from "chemical reaction" type arguments. Specifically, each of the fundamental processes can be likened to a chemical reaction. In chemical terms process (a) becomes

$$(\text{electron}) + \begin{pmatrix} \text{empty} \\ \text{R–G center} \end{pmatrix} \rightarrow \begin{pmatrix} \text{filled} \\ \text{R–G center} \end{pmatrix} \tag{5.3}$$

According to the experimentally verified *rate law* from chemistry, an irreducible chemical reaction of the form $A + B \rightarrow C$ proceeds at a rate given by

$$\text{rate} = (\text{constant})[A][B] \tag{5.4}$$

where $[A]$ and $[B]$ are the concentrations of the reacting components. For process (a) the reacting components are electrons and empty R–G centers; the corresponding concentrations are n and p_T, respectively. Having arrived at the same result is actually not all that surprising because the rate law from chemistry also follows from collision-theory considerations.

Turning to process (b) and invoking a parallel set of arguments, we anticipate a process rate which is directly proportional to the product of the filled R–G center concentration and the concentration of empty conduction band states; i.e.,

$$\left.\frac{\partial n}{\partial t}\right|_{(b)} = (\text{constant}) \begin{pmatrix} \text{concentration of empty} \\ \text{conduction band states} \end{pmatrix} (n_T) \tag{5.5}$$

If the semiconductor is assumed to be nondegenerate, however, the vast majority of conduction band states will be empty at all times. For a nondegenerate semiconductor, then, the concentration of empty conduction band states will be essentially constant and may be incorporated into the process-rate proportionality constant. Thus

$$\left.\frac{\partial n}{\partial t}\right|_{(b)} = e_n n_T \tag{5.6}$$

where

$$e_n \ldots \text{the electron emission coefficient (1/sec)}$$

is again a positive-definite constant. Here the process rate is positive because electron emission always acts to increase the number of electrons in the conduction band.

Analogous arguments can be applied of course to processes (c) and (d). One readily deduces

$$\left.\frac{\partial p}{\partial t}\right|_{(c)} = -c_p n_T p \tag{5.7}$$

and, for a nondegenerate semiconductor,

$$\left.\frac{\partial p}{\partial t}\right|_{(d)} = e_p p_T \tag{5.8}$$

where c_p and e_p are the hole capture and emission coefficients, respectively.

Finally, substituting the fundamental process-rate expressions into Eqs. (5.1), we conclude that

$$r_N \equiv -\left.\frac{\partial n}{\partial t}\right|_{R-G} = c_n p_T n - e_n n_T \tag{5.9a}$$

$$r_P \equiv -\left.\frac{\partial p}{\partial t}\right|_{R-G} = c_p n_T p - e_p p_T \tag{5.9b}$$

In writing down Eqs. (5.9) we have also introduced a more compact net recombination-rate notation. The *net electron and hole recombination rates*, r_N and r_P, are of course positive if recombination is dominant and negative if generation is dominant. Although potentially confusing, this notation does find widespread usage and will be employed subsequently herein. Eqs. (5.9) themselves are very general relationships, applicable in almost any conceivable situation; nondegeneracy is the only limiting restriction. One typically encounters the direct use of these equations in more complex problems (e.g., transient analyses) and in the description of experiments designed to measure the capture and/or emission coefficients.

5.2.3   The Equilibrium Simplification

An intrinsically simpler form of Eqs. (5.9) can be established by invoking the *Principle of Detailed Balance*. Notably, the requirement of detailed balance under equilibrium conditions leads to an interrelationship between the capture and emission coefficients. The statement of the cited principle is as follows:

> *Principle of Detailed Balance.* Under equilibrium conditions each *fundamental* process and its inverse must *self-balance* independent of any other process that may be occurring inside the material.

A corollary of the stated principle provides an excellent definition of the equilibrium state—namely, equilibrium is the special system state where each fundamental process and its inverse self-balance.

When applied to the R–G center interaction, detailed balance requires fundamental process (a) to self-balance with its inverse process (b), and fundamental process (c) to self-balance with its inverse process (d). Consequently,

$$\left.\begin{array}{l} r_N = 0 \\ r_P = 0 \end{array}\right\} \text{ under equilibrium conditions} \qquad \begin{array}{l} \text{(5.10a)} \\ \text{(5.10b)} \end{array}$$

The zero net recombination of carriers under equilibrium conditions forms the mathematical basis for interrelating the emission and capture coefficients. Substituting $r_N = r_P = 0$ into Eqs. (5.9), solving for the emission coefficients, and introducing the subscript "0" to emphasize that all quantities are to be evaluated under equilibrium conditions, one obtains

$$e_{n0} = \frac{c_{n0} p_{T0} n_0}{n_{T0}} = c_{n0} n_1 \qquad \text{(5.11a)}$$

and

$$e_{p0} = \frac{c_{p0} n_{T0} p_0}{p_{T0}} = c_{p0} p_1 \qquad \text{(5.11b)}$$

where

$$\left.\begin{array}{l} n_1 \equiv \dfrac{p_{T0} n_0}{n_{T0}} \\[3mm] p_1 \equiv \dfrac{n_{T0} p_0}{p_{T0}} \end{array}\right\} \begin{array}{l} \text{computable} \\ \text{constants} \end{array} \qquad \begin{array}{l} \text{(5.12a)} \\[3mm] \text{(5.12b)} \end{array}$$

It is next assumed that the emission and capture coefficients all remain approximately equal to their equilibrium values under nonequilibrium conditions; i.e.,

$$e_n \simeq e_{n0} = c_{n0} n_1 \simeq c_n n_1 \qquad \text{(5.13a)}$$

and

$$e_p \simeq e_{p0} = c_{p0} p_1 \simeq c_p p_1 \qquad \text{(5.13b)}$$

Eliminating the emission coefficients in Eqs. (5.9) using Eqs. (5.13) then yields

$$r_N = -\left.\frac{\partial n}{\partial t}\right|_{R-G} = c_n(p_T n - n_T n_1) \qquad \text{(5.14a)}$$

$$r_P = -\left.\frac{\partial p}{\partial t}\right|_{R-G} = c_p(n_T p - p_T p_1) \qquad \text{(5.14b)}$$

Two comments are in order concerning the Eq. (5.14) results. First, like Eqs. (5.9), the Eq. (5.14) results enjoy wide-ranging applicability. There is of course the added assumption that the emission and capture coefficients (actually, the emission to capture coefficient ratios) remain fixed at their equilibrium values. It is difficult to assess the precise limitations imposed by this assumption, although in situations involving large deviations from equilibrium the validity of the equations is certainly open to question.

The second comment addresses the "simplified" nature of the results. At first glance it would appear that we have merely replaced two system parameters (e_n and e_p) with two new system parameters (n_1 and p_1). The emission coefficients are indeed system parameters that must be determined experimentally; n_1 and p_1, however, are computable constants. To facilitate the computation of n_1 and p_1, note that

$$n_1 = \frac{p_{T0}n_0}{n_{T0}} = \left(\frac{N_T - n_{T0}}{n_{T0}}\right)n_0 = \left(\frac{N_T}{n_{T0}} - 1\right)n_0 \tag{5.15}$$

and for a nondegenerate semiconductor

$$n_0 = n_i e^{(E_F - E_i)/kT} \tag{5.16}$$
$$\text{(Same as 4.57a)}$$

Moreover, referring to the n_T definition and Eq. (4.68), one finds

$$\frac{n_{T0}}{N_T} = \frac{1}{1 + e^{(E_T' - E_F)/kT}} \tag{5.17}$$

where $E_T' = E_T \pm kT \ln g_T$. The ($+$) is used if the R–G centers are acceptor-like and the ($-$) if the centers are donor-like. g_T is the degeneracy factor introduced in Chapter 4. Combining Eqs. (5.15) through (5.17) then gives

$$n_1 = n_i e^{(E_T' - E_i)/kT} \tag{5.18a}$$

Likewise

$$p_1 = n_i e^{(E_i - E_T')/kT} \tag{5.18b}$$

Assuming E_T and g_T are known, n_1 and p_1 are readily computed from Eqs. (5.18). For a quick approximate evaluation of these parameters, use can be made of the fact that $n_1 = n_0$ and $p_1 = p_0$ if the Fermi level is positioned such that $E_F = E_T'$. For example, we know that $n_0 = p_0 = n_i$ if $E_F = E_i$ and $n_0 > n_i$, $p_0 < n_i$ if E_F is positioned above midgap. Thus if E_T' is positioned near midgap we analogously conclude $n_1 \simeq p_1 \simeq n_i$ without referring to Eqs. (5.18). Likewise, if E_T' is positioned above midgap, $n_1 > n_i$ and $p_1 < n_i$. Also note that n_1 and p_1 obey the np product relationship: $n_1 p_1 = n_i^2$.

5.2.4 Steady-State Relationship

In the vast majority of device problems the analysis is performed assuming that the device is being operated under steady-state or quasi-steady-state[†] conditions. As we will see, the expressions for the net recombination rates take on a much more tractable form under the cited conditions.

Our first task will be to identify what goes on inside a semiconductor under steady-state conditions and to distinguish between the superficially similar equilibrium and steady states. In both the equilibrium and steady states the average values of all macroscopic observables within a system are constant with time—that is, dn/dt, dp/dt, $d\mathcal{E}/dt$, dn_T/dt, etc. are all zero. Under equilibrium conditions the static situation is maintained by the self-balancing of each fundamental process and its inverse. Under steady-state conditions, on the other hand, the status quo is maintained by a *trade-off* between processes. This difference is nicely illustrated in Fig. 5.6, where the envisioned activity inside a small Δx section of a semiconductor is depicted under equilibrium and steady-state conditions. Please note from Fig. 5.6 that the steady-state net recombination rates are characteristically nonzero.

Although the net recombination rates do not vanish under steady-state conditions, there is nevertheless a readily established interrelationship between the net rates. Since n_T does not change with time, and assuming n_T can only change via the R–G center interaction, one can write

$$\frac{dn_T}{dt} = -\frac{\partial n}{\partial t}\bigg|_{R-G} + \frac{\partial p}{\partial t}\bigg|_{R-G} = r_N - r_P = 0 \tag{5.19}$$

or

$$r_N = r_P \quad \ldots \text{under steady-state conditions} \tag{5.20}$$

The equal creation or annihilation of holes and electrons under steady-state conditions in turn fixes n_T for a given n and p. Specifically, equating the right-hand sides of Eqs. (5.14a) and (5.14b), remembering $p_T = N_T - n_T$, and solving for n_T, one obtains

$$n_T = \frac{c_n N_T n + c_p N_T p_1}{c_n(n + n_1) + c_p(p + p_1)} \quad \text{(steady-state)} \tag{5.21}$$

[†]The quasi-steady-state or quasistatic assumption is invoked quite often in performing transient analyses where the rate of change of system variables such as n, p, $\mathcal{E}$, etc. is slow compared to the rates of the dominant fundamental processes occurring inside the material. Under quasi-steady-state conditions, the instantaneous state of the system may be considered to be a progression of steady states, and steady-state relationships can be used to describe accurately the state of the system at any instant.

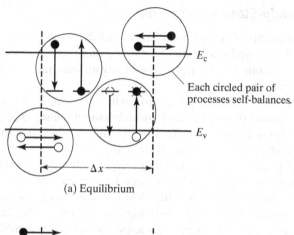

(a) Equilibrium

Each circled pair of
processes self-balances.

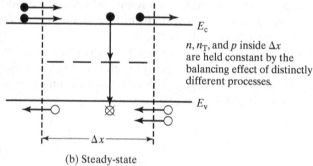

(b) Steady-state

n, n_T, and p inside Δx
are held constant by the
balancing effect of distinctly
different processes.

Figure 5.6   Conceptualization of activity inside a small Δx section of a semiconductor under equilibrium and steady-state conditions.

The Eq. (5.21) n_T expression can next be used to eliminate n_T (and p_T) in either Eq. (5.14a) or Eq. (5.14b). After a bit of manipulation which makes use of the fact that $n_1 p_1 = n_i^2$, we arrive at the result

$$R \equiv r_N = r_P = \frac{np - n_i^2}{\dfrac{1}{c_p N_T}(n + n_1) + \dfrac{1}{c_n N_T}(p + p_1)} \tag{5.22}$$

where the symbol R has been introduced to identify the net steady-state recombination rate. Finally, $1/c_n N_T$ and $1/c_p N_T$ have units of time (seconds) and it is therefore reasonable to additionally introduce the time constants

$$\tau_n = \frac{1}{c_n N_T} \qquad \text{...the electron minority carrier lifetime} \tag{5.23a}$$

and

$$\tau_p = \frac{1}{c_p N_T} \qquad \ldots \text{the hole minority carrier lifetime} \qquad (5.23b)$$

which when substituted into Eq. (5.22) yields

$$R = \frac{np - n_i^2}{\tau_p(n + n_1) + \tau_n(p + p_1)} \qquad (5.24)$$

Equation (5.24) is an extremely important result that is encountered again and again in the device literature. It should be emphasized that the R-expression applies to any steady-state situation and gives the net recombination rate for *both* electrons and holes. The τ's, introduced as a mathematical expedient, are important material parameters and are to be interpreted as the average time an excess minority carrier will live in a sea of majority carriers. This interpretation follows from experiments where an *n*- or *p*-type semiconductor is weakly illuminated with carrier-generating light, the light is extinguished, and the subsequent decay back to equilibrium is monitored as a function of time. Under ideal conditions the time constant for the decay, equal to the average time required to eliminate the excess carriers created by the light, is just τ_n for a *p*-type semiconductor and τ_p for an *n*-type semiconductor. Note that the τ's, simply referred to as the minority carrier lifetimes for identification purposes, vary inversely with the R–G center concentration, but are *explicitly independent* of the doping concentration. Although potentially computable constants, the τ's are routinely deduced directly from experimental measurements. We will have more to say about the minority carrier lifetimes in Section 5.4.

5.2.5 Specialized Steady-State Relationships

The expression for the net steady-state recombination rate can be drastically simplified under certain conditions. In what follows we establish the most widely utilized recombination rate relationships—namely, the reduced steady-state expressions valid (1) under low-level injection conditions and (2) when the semiconductor is depleted of carriers.

Low Level Injection

The level of injection specifies the relative magnitude of changes in the carrier concentrations resulting from a perturbation. Low level injection is said to exist if the changes in the carrier concentrations are much *less* than the majority carrier concentration under equilibrium conditions. Conversely, high level injection exists if the changes are much *greater* than the equilibrium concentration of majority carriers. Mathematically, if n and p are the carrier concentrations under arbitrary conditions, n_0 and p_0 the equilibrium carrier concentrations, and $\boxed{\Delta n = n - n_0}$ and $\boxed{\Delta p = p - p_0}$ the changes in the carrier concentrations from their equilibrium values, then *low level injection implies*

$$\Delta n, \Delta p \ll n_0 \quad (n \simeq n_0) \qquad \ldots \text{in an } n\text{-type material}$$

$$\Delta n, \Delta p \ll p_0 \quad (p \simeq p_0) \qquad \ldots \text{in a } p\text{-type material}$$

Note that low level injection may also be viewed as a situation where the majority carrier concentration remains essentially unperturbed.

To achieve the desired R-expression simplification it is also necessary to make certain gross assumptions about the R–G center parameters. Specifically, we assume that the dominant R–G centers introduce an E_T' level fairly close to midgap so that $n_1 \simeq p_1 \simeq n_i$, and that the R–G center concentration is sufficiently low so that $\Delta n \simeq \Delta p$. Moreover, τ_n and τ_p are taken to differ by no more than a few orders of magnitude. As it turns out, these are reasonable assumptions consistent with the experimentally observed properties of the dominant R–G centers in actual semiconductors.

Proceeding with the simplification, let us first substitute $n = n_0 + \Delta n$ and $p = p_0 + \Delta p$ into Eq. (5.24):

$$R = \frac{(n_0 + \Delta n)(p_0 + \Delta p) - n_i^2}{\tau_p(n_0 + \Delta n + n_1) + \tau_n(p_0 + \Delta p + p_1)} \tag{5.25a}$$

$$\simeq \frac{n_0 p_0 + n_0 \Delta p + p_0 \Delta p + (\Delta p)^2 - n_i^2}{\tau_p(n_0 + \Delta p + n_i) + \tau_n(p_0 + \Delta p + n_i)} \tag{5.25b}$$

In the latter form of Eq. (5.25) we have set $\Delta n = \Delta p$ and $n_1 = p_1 = n_i$ in accordance with previously stated assumptions. Eq. (5.25b) is valid of course for either n- or p-type material. For illustrative purposes let us assume the semiconductor to be n-type. Examining the numerator on the right-hand side of Eq. (5.25b) we note that

$$n_0 p_0 = n_i^2 \rightarrow \text{cancels} - n_i^2$$

$$n_0 \Delta p \gg p_0 \Delta p \quad (n_0 \gg p_0)$$

$$n_0 \Delta p \gg \Delta p^2 \quad (n_0 \gg \Delta p)$$

All but the $n_0 \Delta p$ term may be neglected.

In a like manner, examining the denominator, we note that

$$\tau_p(n_0 + \Delta p + n_i) \simeq \tau_p n_0 \quad (n_0 \gg \Delta p, n_0 \gg n_i)$$

$$\tau_p n_0 \gg \tau_n(p_0 + \Delta p + n_i) \quad (n_0 \gg p_0 + \Delta p + n_i; \quad \tau_n \sim \tau_p)$$

All but the $\tau_p n_0$ term may be neglected.

We therefore arrive at the drastically simplified result:

$$R = \frac{\Delta p}{\tau_p} \quad \dots n\text{-type material} \tag{5.26a}$$

$$R = \frac{\Delta n}{\tau_n} \quad \dots p\text{-type material} \tag{5.26b}$$

The recombination rate expressions obtained here are identical to the "standard case" expressions found in introductory texts and used extensively in device analyses. Eq. (5.25b), we should point out, is valid for any level of injection and, as is readily verified, simplifies to $R = \Delta p/(\tau_n + \tau_p)$ under high level injection conditions where $\Delta n = \Delta p \gg n_0$ or p_0.

R–G Depletion Region

The simplified expression for the net recombination rate inside an R–G depletion region is another special-case result that is encountered quite often in device analyses. An R–G depletion region is formally defined to be a semiconductor volume where $n \ll n_1$ and (simultaneously) $p \ll p_1$. Since $np \ll n_1 p_1 = n_i^2$, a deficit of carriers always exists inside the envisioned depletion region.

Before continuing, it is important to carefully distinguish between an R–G depletion region and the "electrostatic" depletion region encountered in pn-junction analyses. In developing approximate expressions for the electrostatic variables (ρ, $\mathcal{E}$, V) in the pn-junction analysis, it is common practice to assume that the carrier concentrations are negligible compared to the net doping concentration over a width W about the metallurgical junction. This is the well-known depletion approximation and W is the width of the electrostatic depletion region. Unfortunately, the cited terminology can be somewhat misleading. The existence of an electrostatic depletion region merely requires n and p to be small compared to $|N_D - N_A|$; it does not imply the existence of a carrier deficit ($n < n_0$; $p < p_0$) within the region. When a pn-junction is zero biased an electrostatic depletion region exists inside the structure, but $n = n_0$, $p = p_0$, $np = n_i^2$ everywhere because equilibrium conditions prevail. Moreover, when the junction is forward biased there is actually an excess of carriers ($n > n_0$, $p > p_0$) in the electrostatic depletion region. A carrier deficit and associated R–G depletion region are created inside the electrostatic depletion region in a pn-junction only under reverse bias conditions. The width of the R–G depletion region is always smaller than W, but approaches W at large reverse biases.

With $n \ll n_1$ and $p \ll p_1$ in Eq. (5.24), we obtain by inspection,

$$R \simeq -\frac{n_i^2}{\tau_p n_1 + \tau_n p_1} \tag{5.27}$$

or

$$\boxed{G \equiv -R = \frac{n_i}{\tau_g}} \qquad \text{...in an R–G depletion region} \qquad (5.28)$$

where

$$\tau_g \equiv \tau_p(n_1/n_i) + \tau_n(p_1/n_i) \qquad (5.29a)$$

$$= \tau_p + \tau_n \qquad \text{...if } E_T' = E_i \qquad (5.29b)$$

A negative R, of course, indicates that a net generation of carriers is taking place within the depletion region. Hence the introduction of the net generation rate symbol, $G \equiv -R$, and the use of the subscript "g" to identify the *generation lifetime*, τ_g.

5.2.6   Physical View of Carrier Capture

We conclude the discussion of bulk recombination–generation statistics by presenting an alternative derivation of Eq. (5.2) based on a spatially oriented view of the capture process. This derivation has been included because it provides additional insight into the capture process while simultaneously introducing the very useful capture cross section concept.

The real-space visualization of electron capture at an R–G center is shown in Fig. 5.7(a). In this idealized view of the capture process, empty R–G centers are modeled as spheres randomly distributed about the semiconductor volume. Filled R–G centers are thought of as fixed dots and electrons as moving dots. The rather erratic path of the electron as it moves through the semiconductor is caused by collisions with vibrating lattice atoms and ionized impurity atoms. In the course of its travels an electron is considered to have been captured if it penetrates the sphere surrounding an empty R–G center site.

From the qualitative description of the capture process we expect the electron velocity to be a factor in setting the capture rate. Clearly, the greater the distance traveled by the electron per second, the greater the likelihood of electron capture within a given period of time. Now, it can be established in a relatively straightforward manner (see Problem 4.10) that the average kinetic energy of electrons in the conduction band of a nondegenerate semiconductor is $(3/2)kT$ under equilibrium conditions. Thus, the *thermal velocity* or average velocity under equilibrium conditions, v_{th}, is given by

$$\frac{1}{2} m^* v_{th}^2 = \frac{3}{2} kT \qquad (5.30)$$

or

$$v_{th} = \sqrt{3kT/m^*} \simeq 10^7 \text{ cm/sec at 300 K}^\dagger \qquad (5.31)$$

---

†Because of uncertainties in the effective mass to be employed, it is standard practice in R–G computations to use the free electron mass in calculating v_{th}. This applies to both electrons and holes.

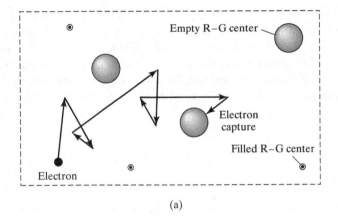

(a)

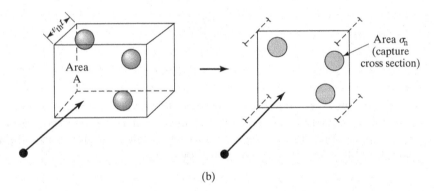

(b)

Figure 5.7 (a) Real-space visualization of electron capture at an R–G center. (b) Construct employed in determining the capture rate. Left: volume through which an electron passes in a time t. Right: effective recombination plane.

Under nonequilibrium conditions there are of course added velocity components. However, the added velocity is typically small compared to v_{th} and, as a general rule, little error is introduced by assuming $v \simeq v_{th}$ in modeling the capture process under arbitrary conditions.

We can now turn to the derivation proper. In a time t (assumed to be small) an electron will travel a distance $v_{th}t$ and will pass through a volume of material equal to $Av_{th}t$, where A is the cross-sectional area of the material normal to the electron's path. In this volume there will be p_T empty R–G centers per cm^3, or a total number of $p_T Av_{th}t$ empty R–G centers. Since the R–G centers are assumed to be randomly distributed, the probability of the electron being captured in the volume can be determined by conceptually moving the centers of all R–G spheres to a single plane in the middle of the volume and noting the fraction of the plane blocked by the R–G centers (see Fig. 5.7(b)). If the area of the plane blocked by a single R–G center is $\sigma_n = \pi r^2$, where r is the radius of the R–G spheres, the total area blocked by empty R–G centers will be $p_T A \sigma_n v_{th} t$. The fraction of the area giving rise to capture will be $p_T A \sigma_n v_{th} t / A$. The probability of electron capture in the volume is then $p_T \sigma_n v_{th} t$, and the capture rate (probability of capture per second) for a single electron is $p_T \sigma_n v_{th} t / t = p_T \sigma_n v_{th}$. Given n electrons per unit volume, the number of electrons/cm^3 captured per second will be $n p_T \sigma_n v_{th}$, or

$$\left.\frac{\partial n}{\partial t}\right|_{(a)} = -\sigma_n v_{th} p_T n \tag{5.32}$$

Equations (5.2) and (5.32) are clearly equivalent if one identifies

$$c_n = \sigma_n v_{th} \tag{5.33a}$$

Analogously

$$c_p = \sigma_p v_{th} \tag{5.33b}$$

σ_n and σ_p, the electron and hole *capture cross sections*, are often used to gauge the relative effectiveness of R–G centers in capturing carriers. In fact, because of their "intuitive" appeal, the capture cross sections find a much greater utilization in the device literature than the more basic capture coefficients.

5.3 SURFACE RECOMBINATION–GENERATION

5.3.1 Introductory Comments

In many devices under certain conditions, surface recombination–generation can be as important as, or more important than, the "bulk" recombination–generation considered in the preceding section. Whereas bulk R–G takes place at centers spatially distributed throughout the volume of a semiconductor, surface recombination–generation refers to the creation/annihilation of carriers in the near vicinity of a semiconductor surface via the interaction with interfacial traps. Interfacial traps or surface states are functionally equivalent to R–G centers localized at the surface of a material. Unlike bulk R–G centers, however, interfacial traps are typically found to be continuously distributed in energy throughout the semiconductor band gap.

As pictured in Fig. 5.8, the same fundamental processes that occur in the semiconductor bulk also occur at the semiconductor surface. Electrons and holes can be captured at surface centers; electrons and holes can be emitted from surface centers. From the energy band description alone one might expect additional transitions to occur between surface centers at different energies. However, given realistic interfacial-trap densities, these seemingly plausible intercenter transitions are extremely unlikely

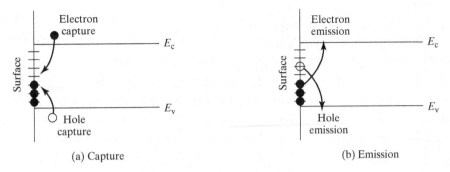

Figure 5.8 Recombination–generation at a semiconductor surface via transitions to and from interfacial traps. (a) Electron and hole capture leading to carrier recombination. (b) Electron and hole emission leading to carrier generation.

because of the spread-out or spatially isolated nature of the centers on the surface plane (see Fig. 5.9).

The very obvious physical similarity between surface and bulk recombination–generation leads to a parallel mathematical description of the processes. This will allow us to establish a number of surface relationships by direct inference from the corresponding bulk result. Nevertheless, there are two major differences:

(1) Because surface states are arranged along a plane in space rather than spread out over a volume, the net recombination rates are logically expressed in terms of carriers removed from a given band per UNIT AREA–second.

(2) Whereas a single level usually dominates bulk recombination–generation, the surface-center interaction routinely involves centers distributed in energy throughout the band gap. Hence, it is necessary to add up or integrate the single-level surface rates over the energy band gap.

5.3.2 General Rate Relationships (Single Level)

It is convenient to initially determine the net recombination rates associated with interface traps at a single energy, and to subsequently modify the results to account for

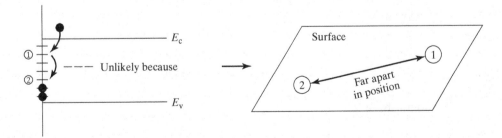

Figure 5.9 Visualization of an intercenter surface transition and a pictorial explanation of why such transitions are highly unlikely.

the distributed nature of the states. To begin the analysis we therefore assume the band gap contains a single energy level, E_{IT}. Adding the subscript s to the corresponding bulk definitions, let us also define:

r_{Ns} ...   Net electron recombination rate at surface centers (that is, the net change in the number of conduction band electrons/cm²-sec due to electron capture and emission at the single level surface centers).

r_{Ps} ...   Net hole recombination rate at the surface centers.

n_{Ts} ...   Filled surface centers/**cm²** at energy E_{IT}.

p_{Ts} ...   Empty surface centers/**cm²** at energy E_{IT}.

N_{Ts} ...   Total number of surface states/cm²; $N_{Ts} = n_{Ts} + p_{Ts}$.

n_s ...   Surface electron concentration (number per cm CUBED); $n_s = n|_{surface}$.

p_s ...   Surface hole concentration.

e_{ns}, e_{ps} ...   Surface electron and hole emission coefficients (1/sec).

c_{ns}, c_{ps} ...   Surface electron and hole capture coefficients (cm³/sec).

Given the one-to-one correspondence between physical processes and parametric quantities, we likewise expect a term-by-term correspondence in the expressions for the surface and bulk net recombination rates. For the bulk process we obtained the general relationships

$$r_N = c_n p_T n - e_n n_T$$
$$r_P = c_p n_T p - e_p p_T$$

Thus, for single level surface recombination–generation we conclude by analogy

$$\left[\begin{array}{l} r_{Ns} = c_{ns} p_{Ts} n_s - e_{ns} n_{Ts} \\ r_{Ps} = c_{ps} n_{Ts} p_s - e_{ps} p_{Ts} \end{array} \right.$$

(5.34a)

(5.34b)

Like their bulk counterparts, Eqs. (5.34) are very general relationships, with nondegeneracy being the only limiting restriction.

In the bulk analysis we next invoked detailed balance to obtain the simplified general relationships

$$r_N = c_n(p_T n - n_T n_i)$$
$$r_P = c_p(n_T p - p_T p_i)$$

Analogously, therefore

$$\left[\begin{array}{l} r_{Ns} = c_{ns}(p_{Ts} n_s - n_{Ts} n_{is}) \\ r_{Ps} = c_{ps}(n_{Ts} p_s - p_{Ts} p_{is}) \end{array} \right.$$

(5.35a)

(5.35b)

where

$$e_{ns} = c_{ns}n_{1s} \tag{5.36a}$$

$$e_{ps} = c_{ps}p_{1s} \tag{5.36b}$$

and, taking the surface-center degeneracy factor to be unity,

$$n_{1s} = n_i e^{(E_{TT}-E_i)/kT} \tag{5.37a}$$

$$p_{1s} = n_i e^{(E_i-E_{TT})/kT} \tag{5.37b}$$

5.3.3 Steady-State Relationships

Single Level

Under steady-state conditions, as in the bulk,

$$r_{Ns} = r_{Ps} \equiv R_s \tag{5.38}$$

if the filled-state population of interfacial traps at E_{IT} is assumed to change exclusively via thermal band-to-trap interactions. Equating the right-hand sides of Eqs. (5.35a) and (5.35b), one obtains

$$n_{Ts} = \frac{c_{ns}N_{Ts}n_s + c_{ps}N_{Ts}p_{1s}}{c_{ns}(n_s + n_{1s}) + c_{ps}(p_s + p_{1s})} \quad \left(\begin{array}{c} \text{steady} \\ \text{state} \end{array} \right) \tag{5.39}$$

Equation (5.39) can then be used to eliminate n_{Ts} (and p_{Ts}) in either Eq. (5.35a) or Eq. (5.35b), yielding

$$\left[R_s = \frac{n_s p_s - n_i^2}{\dfrac{1}{c_{ps}N_{Ts}}(n_s + n_{1s}) + \dfrac{1}{c_{ns}N_{Ts}}(p_s + p_{1s})} \right] \tag{5.40}$$

Again, Eq. (5.40) is seen to be the direct surface analog of the corresponding bulk result [Eq. (5.22)]. Please note, however, that $1/c_{ns}N_{Ts}$ and $1/c_{ps}N_{Ts}$ are NOT time constants. In fact, $c_{ns}N_{Ts} \equiv s_n$ and $c_{ps}N_{Ts} \equiv s_p$ have units of a velocity, cm/sec, and are, respectively, the (single level) surface recombination velocities for electrons and holes. Conceptually, the recombination of excess carriers at a surface causes a flow of carriers toward the surface. Provided that low level injection conditions prevail and the surface bands are flat ($\mathscr{E}|_{surface} = 0$), the velocities at which the excess carriers flow into the surface will be s_n and s_p, respectively, in p- and n-type semiconductors containing a single

E_{IT} level. Because the single-level case is of little practical interest, the s_n and s_p velocities as defined above are unlikely to be encountered in the device literature. Nevertheless, it is commonplace to encounter the symbol s and an appropriately defined surface recombination (or generation) velocity in surface analyses. Functionally, the s in a surface R–G analysis replaces the τ in the corresponding bulk R–G analysis as the material constant characterizing the net carrier action.

Multi-Level

As already noted, surface centers are typically found to be continuously distributed in energy throughout the semiconductor band gap. The net recombination rates associated with the individual centers in the distribution must be added together to obtain the overall net recombination rate. A simple addition of rates is possible, we should interject, because the centers at different energies are noninteracting (i.e., as previously described, inter-center transitions are extremely unlikely). The task at hand is to appropriately modify the single-level result to obtain the net recombination rate associated with a continuous distribution of noninteracting surface centers. Although we specifically consider the steady-state case, a similar modification procedure can be readily applied to any of the foregoing single-level results.

Let $D_{\text{IT}}(E)$ be the density of interfacial traps (traps per cm²-eV) at an arbitrarily chosen energy E $(E_v \leq E \leq E_c)$. $D_{\text{IT}}(E)dE$ will then be the number of interfacial traps per cm² with energies between E and $E + dE$. Associating $D_{\text{IT}}(E)dE$ with N_{Ts} in the single-level relationship $[N_{Ts} \rightarrow D_{\text{IT}}(E)dE$ in Eq. (5.40)] and recognizing that these states provide an incremental contribution (dR_s) to the overall net recombination rate when there is a distribution of states, one deduces

$$dR_s = \frac{n_s p_s - n_i^2}{(n_s + n_{1s})/c_{ps} + (p_s + p_{1s})/c_{ns}} D_{\text{IT}}(E)dE \tag{5.41}$$

$$= \text{net recombination rate associated with centers between } E \text{ and } E + dE$$

Integrating over all band gap energies then yields

$$R_s = \int_{E_v}^{E_c} \frac{n_s p_s - n_i^2}{(n_s + n_{1s})/c_{ps} + (p_s + p_{1s})/c_{ns}} D_{\text{IT}}(E)dE \tag{5.42}$$

In utilizing the above relationship it must be remembered that all of the trap parameters can vary with energy. The anticipated variation of $D_{\text{IT}}(E)$ with energy is of course noted explicitly. With the integration variable E replacing E_{IT} in Eqs. (5.37), n_{1s} and p_{1s} are also seen to be functions of energy. In fact, n_{1s} and p_{1s} are exponential functions of energy. Even c_{ns} and c_{ps} can vary with the trap energy across the band gap. Like $D_{\text{IT}}(E)$, however, $c_{ns}(E)$ and $c_{ps}(E)$ must be determined from experimental measurements.

5.3.4 Specialized Steady-State Relationships

To conclude the discussion of surface recombination–generation we consider two special cases of practical interest that give rise to simplified R_s relationships. The special-case analyses, treating (1) low level injection when the bands are flat and (2) a depleted ($n_s \rightarrow 0$, $p_s \rightarrow 0$) surface, nicely illustrate simplification procedures and are the surface analog of the bulk analyses presented in Subsection 5.2.5.

Low Level Injection/Flat Band

We assume the semiconductor under analysis is n-type, the energy bands are flat at the surface ($\mathcal{E}|_{surface} = 0$, implying $n_{s0} = N_D$), and low level injection conditions prevail ($\Delta n_s = \Delta p_s \ll n_{s0}$). c_{ns} and c_{ps} are also taken to be comparable in magnitude. Under the stated conditions and introducing $n_s = n_{s0} + \Delta p_s$, $p_s = p_{s0} + \Delta p_s$, we find

$$n_s p_s - n_i^2 \simeq n_{s0} \Delta p_s \tag{5.43}$$

and

$$(n_s + n_{1s})/c_{ps} + (p_s + p_{1s})/c_{ns} \simeq (n_{s0} + n_{1s})/c_{ps} + p_{1s}/c_{ns} \tag{5.44}$$

Thus Eq. (5.42) simplifies to

$$R_s \simeq \left[\int_{E_v}^{E_c} \frac{c_{ps} D_{IT}}{1 + \dfrac{n_{1s}}{n_{s0}} + \dfrac{c_{ps}}{c_{ns}} \dfrac{p_{1s}}{n_{s0}}} \, dE \right] \Delta p_s \tag{5.45}$$

For a given set of trap parameters the integral in Eq. (5.45) is a system constant and must have the dimensions of a velocity. Logically taking this integral to be a surface recombination velocity, we can therefore write

$$\boxed{R_s = s_p \Delta p_s} \qquad \ldots n\text{-type material} \tag{5.46}$$

where

$$s_p \equiv \int_{E_v}^{E_c} \frac{c_{ps} D_{IT}}{1 + \dfrac{n_{1s}}{n_{s0}} + \dfrac{c_{ps}}{c_{ns}} \dfrac{p_{1s}}{n_{s0}}} \, dE \tag{5.47}$$

An analogous result is obtained for p-type material.

The simplification procedure can be carried one step further if more limiting assumptions are made concerning the trap parameters. Notably, D_{IT} and the capture coefficients are often assumed to be approximately constant (energy-independent) over the middle portion of the band gap. Indeed, a certain amount of experimental Si data tends to support this assumption[4,5]. Let us pursue the implications of the assumption. Referring to Fig. 5.10(a), which provides a sketch of the denominator in the s_p integrand versus energy, note

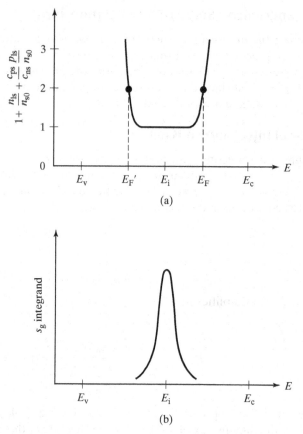

Figure 5.10   Observations related to special-case simplification of the R_s relationship. Sketches of (a) the denominator in the s_p integrand [Eq. (5.47)] and (b) the s_g integrand [Eq. (5.52b)] as a function of band gap energy.

$$1 + \frac{n_{1s}}{n_{s0}} + \frac{c_{ps}}{c_{ns}}\frac{p_{1s}}{n_{s0}} = 1 + \frac{n_i}{N_D}\left[e^{(E-E_i)/kT} + \frac{c_{ps}}{c_{ns}}e^{(E_i-E)/kT}\right] \qquad (5.48)$$

$$\simeq 1 \qquad \text{...if } E_F{}' \leq E \leq E_F \qquad (5.49a)$$

$$\rightarrow \infty \qquad \text{...for } E < E_F{}' \text{ and } E > E_F \qquad (5.49b)$$

$E_F{}'$ being the energy in the band gap where

$$\frac{n_i}{N_D}\frac{c_{ps}}{c_{ns}}e^{(E_i-E)/kT} = 1 \qquad (5.50)$$

In other words, the denominator of the s_p integrand is approximately unity in the midgap region where D_{IT} and the capture coefficients are assumed to be approximately constant. Outside this range the denominator becomes large and the contribution to the overall integral is small. Thus, if D_{IT} and the capture coefficients are taken to be constant over the midgap range,

$$s_p \simeq \int_{E_F'}^{E_F} c_{ps} D_{IT} \mathrm{d}E \simeq \underbrace{c_{ps} D_{IT}} (E_F - E_F')$$ (5.51)

evaluated at midgap

The Eq. (5.51) result is useful for estimating s_p. It also has the same general form as the single-level surface recombination velocity ($s_p = c_{ps} N_{Ts}$), thereby adding credibility to our use of the s_p symbol in the present context.

Before concluding, a comment is in order concerning the earlier $R_s = s_p \Delta p_s$ result. As is clearly evident from the analysis, this result is only valid under rather restrictive conditions. Nevertheless, in analyzing solar cells, photodetectors, and other photodevices, it is all but universally assumed that one can write $R_s = s \Delta p_s$ (or $R_s = s \Delta n_s$ for p-type material) UNDER ARBITRARY CONDITIONS, with the "surface recombination velocity" s being treated as a system constant. Admittedly, one can introduce a "generalized" surface recombination velocity, $s \equiv R_s[\text{Eq. (5.42)}]/\Delta p_s$ for n-type material and $s \equiv R_s[\text{Eq. (5.42)}]/\Delta n_s$ for p-type material. However, the s thus defined is not necessarily a system constant. The generalized s will vary with the level of injection and the amount of band bending. Moreover, under certain conditions s will even be a function of the perturbed carrier concentrations. Nonetheless, the somewhat questionable practice of treating s as a system constant (often a matter of expediency) persists, and it must be acknowledged.

Depleted Surface

If non-equilibrium conditions exist such that both $n_s \to 0$ and $p_s \to 0$ at the surface of a semiconductor, R_s by inspection reduces to

$$R_s = \int_{E_v}^{E_c} \frac{-n_i^2}{n_{1s}/c_{ps} + p_{1s}/c_{ns}} D_{IT}(E) \mathrm{d}E$$ (5.52a)

$$= -\left[\int_{E_v}^{E_c} \frac{c_{ns} c_{ps} D_{IT} \mathrm{d}E}{c_{ns} e^{(E-E_i)/kT} + c_{ps} e^{(E_i-E)/kT}} \right] n_i$$ (5.52b)

The integral in Eq. (5.52b) is a system constant with the dimensions of a velocity and is commonly referred to as the *surface generation velocity*, s_g. A negative R_s indicates, of course, that carriers are being generated at the surface. Taking the trap parameters in the Eq. (5.52b) integral to be reasonably well-behaved functions of energy, we find the s_g integrand to be a highly peaked function of E, maximizing at $E \simeq E_i$ if

$c_{ns} = c_{ps}$ (see Fig. 5.10(b)). Given the highly peaked nature of the integrand, little error is introduced by (a) assuming the trap parameters are constant at their midgap values $[D_{IT}(E) = D_{IT}(E_i)$, etc.] and (b) setting the lower and upper integration limits to $-\infty$ and $+\infty$, respectively. A closed-form evaluation of the integral then becomes possible, giving

$$G_s \equiv -R_s = s_g n_i \qquad \left(\begin{array}{c}\text{Depleted} \\ \text{surface}\end{array}\right) \qquad (5.53)$$

$$s_g = \frac{\pi}{2}\sqrt{c_{ns}c_{ps}}\,kTD_{IT} \qquad \qquad \qquad (5.54)$$

where the trap parameters in the classic Eq. (5.54) result are to be evaluated at midgap.

5.4 SUPPLEMENTAL R–G INFORMATION

Collected in this section is a collage of practical R–G center information that is intended to enhance and supplement the preceding theoretical description of the R–G center process.

Multistep Nature of Carrier Capture

During our survey of R–G processes the recombination of carriers at R–G centers was noted to be typically non-radiative, implying that the energy lost in the recombination process gives rise to lattice vibrations or phonons. As noted subsequently, however, a single phonon can only carry away a small amount of energy. One is therefore faced with somewhat of a logical dilemma. If a large number of phonons must simultaneously collide with a carrier for capture to occur, the R–G center recombination process would be extremely unlikely! This dilemma is resolved by a more detailed view of carrier capture at R–G centers that recognizes the multistep nature of the process[6–8]. In the cascade model[7] an electron (or hole) is viewed to be first weakly bound in an excited-state orbit about the R–G center site. As the carrier moves about the R–G center it loses energy in small increments via collisions with the semiconductor lattice. With the sequential loss in energy to phonons, the carrier spirals in toward the R–G center and is ultimately "captured" or bound tightly to the center. In the multiphonon model[8] the carrier is viewed as transferring most of its energy in an initial step that causes a violent lattice vibration in the vicinity of the R–G center. The vibration subsequently damps down to the amplitude of thermal vibrations after a small number of vibrational periods. During the damping the localized energy and momentum are carried away from the R–G center by lattice phonons. A cascade-like model appears to describe the initial portion of capture at ionized R–G centers, while a multiphonon-like model is considered to be applicable for the final portion of capture at ionized centers and for the entire capture process at neutral centers.

Manipulation of N_T

The carrier lifetimes (τ_n and τ_p) within a given material determine the response time of the R–G center interaction. The lifetimes in turn are inversely proportional to N_T, the concentration of the dominant R–G center inside the material. Generally speaking, the dominant R–G center will be an unintentional impurity (or impurity-decorated defect) incorporated into the material during crystal growth or device processing. The resulting lifetimes are often variable and quite unpredictable. Thus procedures have evolved whereby N_T can be manipulated to optimize the characteristics of devices intended for τ-sensitive applications. The lifetimes can be controllably decreased by simply adding known amounts of an efficient R–G center to the semiconductor—e.g., by diffusing Au into Si. To increase the lifetimes, one or more "gettering" steps are included in the device fabrication procedure. "Gettering," or removal of R–G centers from the portion of the semiconductor containing active device junctions, can be accomplished, for example, by diffusing phosphorus into the back side of a Si wafer. During this high-temperature process, R–G centers move about the semiconductor and become trapped in the back-side layer away from the active front surface. For additional gettering information the reader is referred to the device fabrication literature[9].

Selected Bulk Parametric Data

As we have indicated, the observed minority carrier lifetimes can vary dramatically with the quality of the starting semiconductor, the nature and number of device processing steps, and whether or not the R–G center concentration has been intentionally manipulated during device fabrication. Although a single universal value cannot be quoted, it is nevertheless worthwhile to indicate the general range of lifetimes to be expected under certain conditions in an extensively researched semiconductor such as silicon. Specifically, observed Si lifetimes can be grouped into three ranges as summarized in Table 5.1. The longest reported lifetimes, some exceeding 1 msec, are usually derived from device structures which are devoid of active pn-junctions and whose construction involves a minimum of high-temperature processing steps. On the other hand, pn-junction devices are routinely characterized by carrier lifetimes ranging from 1 to 100 μsec. It should be noted that the τ_n and τ_p entries in Table 5.1, which are characteristic of junction devices, were simultaneously measured with a single test structure by varying the level of injection. (The vast majority of lifetime measurements reported in the literature give only τ_g, τ_n in p-type material, or τ_p in n-type material.) Finally, sub-microsecond lifetimes are readily achieved by design in Au-diffused Si structures. Gold introduces two levels into the Si band gap (see Fig. 4.14), but often one of the levels dominates the R–G center interaction. In such instances, as the Table 5.1 entry implies, the Au interaction may be acceptably modeled by single-level statistics.

Table 5.1 Observed Carrier Lifetimes in Si (300 K)

Lifetime Range	Result-Reference	Device Structure
10^{-4}–10^{-2} sec	$\tau_g = 2$ msec [10]	Gettered Metal/SiO$_2$/Si (MOS) capacitors
10^{-6}–10^{-4} sec	$\tau_n = 23.5\ \mu$sec $\tau_p = 1.5\ \mu$sec [11]	pn-junction devices
10^{-8}–10^{-6} sec	$\tau_n = 0.75\ \mu$sec $\tau_p = 0.25\ \mu$sec [11]	Au-diffused pn-junction devices

Table 5.2 Measured Capture Cross Sections in Si (300 K)

R–G Center	Capture Cross Section[†]		Reference
Au	$\sigma_n \simeq 1 \times 10^{-16} \text{cm}^2$	... acceptor level	[12]
	$\sigma_p = 1 \times 10^{-13} \text{cm}^2$		
	$\sigma_n = 6.3 \times 10^{-15} \text{cm}^2$	... donor level	[13]
	$\sigma_p = 2.4 \times 10^{-15} \text{cm}^2$		
Pt	$\sigma_n = 3.2 \times 10^{-14} \text{cm}^2$	... acceptor level	
	$\sigma_p = 2.7 \times 10^{-12} \text{cm}^2$	closest to midgap	[14]
Zn[‡]	$\sigma_n = 2 \times 10^{-16} \text{cm}^2$	... $\text{Zn}^0 + \text{e}^- \rightarrow \text{Zn}^-$	
	$\sigma_p = 6 \times 10^{-15} \text{cm}^2$	... $\text{Zn}^- + \text{h}^+ \rightarrow \text{Zn}^0$	
	$\sigma_n = 5 \times 10^{-18} \text{cm}^2$	... $\text{Zn}^- + \text{e}^- \rightarrow \text{Zn}^{--}$	[15]
	$\sigma_p = 7 \times 10^{-14} \text{cm}^2$	... $\text{Zn}^{--} + \text{h}^+ \rightarrow \text{Zn}^-$	

[†] A $v_{th} = 10^7$ cm/sec was assumed in converting capture coefficients to capture cross sections.

[‡] Observed σ_p capture cross sections were field-dependent. Quoted values are for $\mathscr{E} = 10^4$ V/cm.

Parametric data of a more fundamental nature, including R–G center concentrations, emission rate coefficients, and capture cross sections, can also be found in the device literature. A sampling of capture cross section results is presented in Table 5.2. Since Au in Si is the foremost example of an efficient R–G center, the Au in Si capture cross sections provide a standard of comparison for gauging the effectiveness of other centers. We should note that, with increasingly stringent material requirements in device manufacture, there arose a growing need for more detailed information about R–G centers. This in turn led to the development of routine measurement techniques, notably DLTS[16] for determining fundamental R–G center parameters. Commercial DLTS (Deep Level Transient Spectroscopy) systems can be used to determine energy levels, emission coefficients, trap concentrations, and capture cross sections. The technique is primarily limited to the detection of purposely introduced centers in Si, but can readily detect the higher levels of unintentional R–G centers present in other materials.

Doping Dependence

The minority carrier lifetimes as defined by Eqs. (5.23) are explicitly independent of the acceptor and donor concentrations. However, Auger recombination, with $\tau_{Auger} \propto 1/(\text{carrier concentration})^2$, becomes the dominant recombination mechanism at high doping levels. As a result, the carrier lifetimes in Si exhibit a decrease with increased doping roughly as pictured in Fig. 5.11. In addition, for n-type Si, there is a theoretical prediction and supporting experimental evidence of a decrease in the minority carrier lifetime at doping levels below the onset of significant Auger recombination. This is believed to result from an increase in the defect density, and hence an added N_T, in direct proportion to the N_D doping concentration[17a].

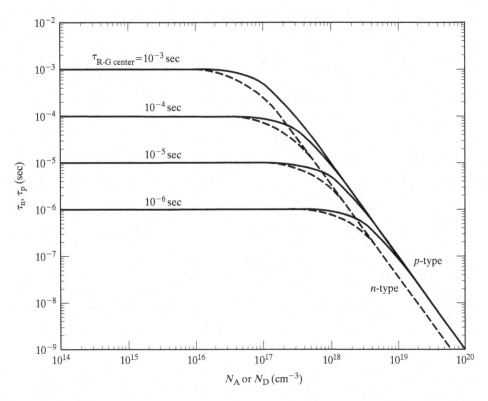

Figure 5.11   Effect of Auger recombination at high doping levels on the carrier lifetimes in Si at 300 K. Lifetimes computed employing the Auger recombination coefficients found in Reference [17b].

Selected Surface Parametric Data (Si/SiO_2)

The surface parametric data to be examined specifically applies to the Si/thermally grown SiO_2 interface. Because of its technological importance, the Si/SiO$_2$ interface has been the subject of an intensive experimental investigation. The available information on the interface state parameters characterizing the oxide-covered Si surface is quite extensive, far exceeding (and in greater detail than) that on all other semiconductor surfaces and interfaces combined. This is not to say that the Si/SiO$_2$ interface is thoroughly characterized. One complication stems from the fact that the surface state parameters are strongly process-dependent. This dependence is nicely illustrated by Fig. 5.12, reproduced from Ref. [18]. Whereas $D_{IT}(E)$ is at a minimum and approximately constant over the midgap region in an "optimally" processed structure, small variations in processing can give rise to interface-state densities which are larger by several orders of magnitude and which exhibit a decidedly different energy dependence. Surface state parameters also vary systematically with the Si surface orientation and are affected by ionizing radiation.

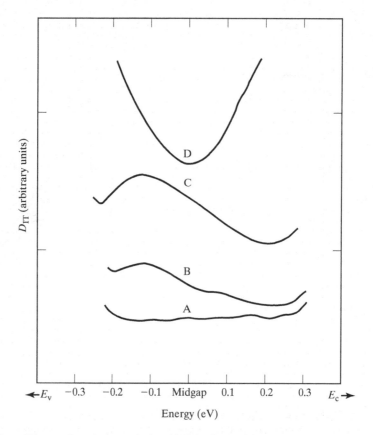

Figure 5.12 Observed variation of the Si/SiO$_2$ interfacial trap density distribution with processing steps immediately following thermal oxidation of the Si surface. The curves shown illustrate general trends. (A) Sample with near-optimum H$_2$ anneal. (B) Sample with nonoptimum H$_2$ anneal. (C) Unannealed sample pulled in dry O$_2$. (D) Unannealed sample pulled in N$_2$. (From Razouk and Deal[18]. Reprinted by permission of the publisher, The Electrochemical Society, Inc.)

The parametric data presented in Fig. 5.13 and Fig. 5.14 are representative of an "optimally" processed, (100)-oriented, thermally oxidized Si surface. The plots are a superposition of the results reported by a number of investigators[4, 5, 19–21, 23]. Given the independent fabrication of test structures, the difficulty of the measurements, and the use of different measurement techniques, the agreement between results is generally quite good.

In examining Fig. 5.13, please note that D_{IT} is approximately constant over the midgap region, with midgap values $\sim 10^{10}$ states/cm^2-eV.[†] Most of the data sets

---

[†]Although 10^{10}/cm^2-eV is a fairly representative value, midgap surface state densities as low as 2×10^9/cm^2-eV have been reported[22].

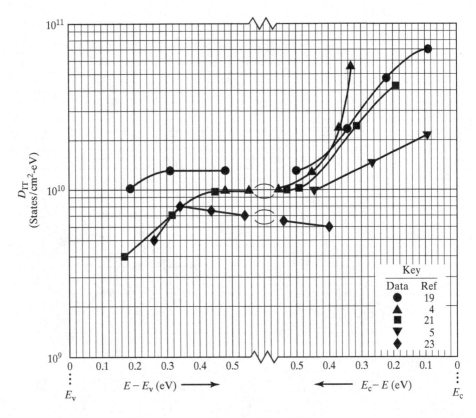

Figure 5.13 Measured Si/SiO$_2$ interfacial trap densities as a function of band gap energy[4, 5, 19, 21, 23]. The data are representative of that derived from an optimally (or near-optimally) processed, (100)-oriented, thermally oxidized Si surface.

likewise show an increase in D_{IT} as one approaches the conduction band edge. Interestingly, a similar increase is not noted near the valence band edge. This, however, may be a function of the measurement techniques (DLTS and Charge Pumping) used to acquire the near-E_v Fig. 5.13 data. Other techniques applied to structures with higher surface state densities have consistently shown an upturn in D_{IT} near the valence band edge. Finally, note in Fig. 5.14 that available results for σ_{ns} and σ_{ps} ($c_{ns} = \sigma_{ns}v_{th}$, $c_{ps} = \sigma_{ps}v_{th}$) are primarily confined respectively to the upper and lower halves of the band gap. σ_{ns} is generally observed to be approximately constant or slowly varying near midgap, while falling off sharply as E approaches E_c. σ_{ps} results, on the other hand, have not exhibited a consistent trend. Measurement complications arise from the interplay between energy, temperature, and surface field dependencies. In addition, it is possible there may be more than one type of interface trap.

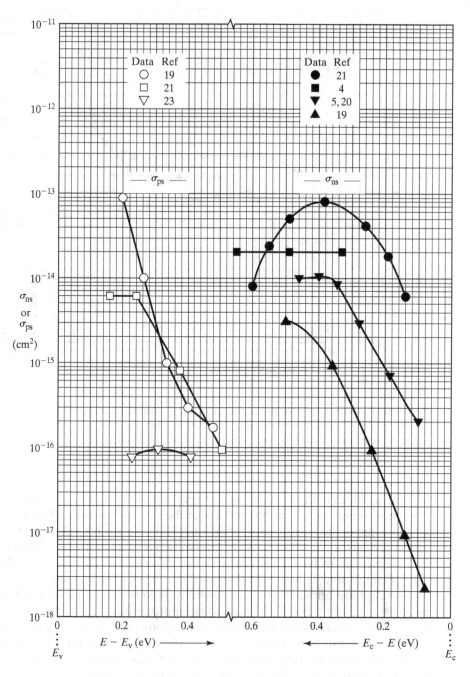

Figure 5.14   Measured electron and hole surface capture cross sections[4,5,19–21,23] characteristic of an optimally processed, (100)-oriented, thermally oxidized Si surface.

REFERENCES

[1] W. Shockley and W. T. Read, Jr., "Statistics of the Recombination of Holes and Electrons," Phys. Rev., *87*, 835 (1952).

[2] R. N. Hall, "Germanium Rectifier Characteristics," Phys. Rev., *83*, 228 (1951).

[3] R. N. Hall, "Electron-Hole Recombination in Germanium," Phys. Rev., *87*, 387 (1952).

[4] J. A. Cooper, Jr. and R. J. Schwartz, "Electrical Characteristics of the SiO_2–Si Interface Near Midgap and in Weak Inversion," Solid-State Electronics, *17*, 641 (1974).

[5] H. Deuling, E. Klausmann, and A. Goetzberger, "Interface States in Si–SiO_2 Interfaces," Solid-State Electronics, *15*, 559 (1972).

[6] H. J. Queisser, "Recombination at Deep Traps," Solid-State Electronics, *21*, 1495 (1980).

[7] M. Lax, "Cascade Capture of Electrons in Solids," Phys. Rev., *119*, 1502 (1960).

[8] C. H. Henry and D. V. Lang, "Nonradiative Capture and Recombination by Multiphonon Emission in GaAs and GaP," Phys. Rev. B, *15*, 989 (Jan., 1977).

[9] For an excellent discussion of gettering processes see J. D. Plummer, M. D. Deal, and P. B. Griffin, *Silicon VLSI Technology, Fundamentals, Practice, and Modeling*, Prentice Hall, Upper Saddle River NJ, 2000.

[10] A. Rohatgi and P. Rai-Choudhury, "Process-Induced Effects on Carrier Lifetime and Defects in Float Zone Silicon," J. Electrochem. Soc., *127*, 1136 (May, 1980).

[11] W. Zimmerman, "Experimental Verification of the Shockley-Read-Hall Recombination Theory in Silicon," Electronics Letts., *9*, 378 (Aug., 1973).

[12] F. Richou, G. Pelous, and D. Lecrosnier, "Thermal Generation of Carriers in Gold-Doped Silicon," J. Appl. Phys., *51*, 6252, (Dec., 1980).

[13] J. M. Fairfield and B. V. Gokhale, "Gold as a Recombination Centre in Silicon," Solid-State Electronics, *8*, 685 (1965).

[14] K. P. Lisiak and A. G. Milnes, "Platinum as a Lifetime-Control Deep Impurity in Silicon," J. Appl. Phys., *46*, 5229 (Dec., 1975).

[15] J. M. Herman III and C. T. Sah, "Thermal Capture of Electrons and Holes at Zinc Centers in Silicon," Solid-State Electronics, *16*, 1133 (1973).

[16] D. V. Lang, "Deep-Level Transient Spectroscopy: A New Method to Characterize Traps in Semiconductors," J. Appl. Phys., *45*, 3023 (July, 1974).

[17a] R. J. Van Overstraeten and R. P. Mertens, "Heavy Doping Effects in Silicon," Solid-State Electronics, *30*, 1077 (1987).

[17b] J. Dziewior and W. Schmid, "Auger Coefficients for Highly Doped and Highly Excited Silicon," Appl. Phys. Lett., *31*, 346 (Sept., 1977).

[18] R. R. Razouk and B. E. Deal, "Dependence of Interface State Density on Silicon Thermal Oxidation Process Variables," J. Electrochem. Soc., *126*, 1573 (Sept., 1979).

[19] T. J. Tredwell and C. R. Viswanathan, "Interface-State Parameter Determination by Deep-Level Transient Spectroscopy," Appl. Phys. Lett., *36*, 462 (March 15, 1980).

[20] T. Katsube, K. Kakimoto, and T. Ikoma, "Temperature and Energy Dependence of Capture Cross Sections at Surface States in Si Metal-Oxide-Semiconductor Diodes Measured by Deep Level Transient Spectroscopy," J. Appl. Phys., *52*, 3504 (May, 1981).

[21] W. D. Eades and R. M. Swanson, "Improvements in the Determination of Interface State Density Using Deep Level Transient Spectroscopy," J. Appl. Phys., 56, 1744 (Sept., 1984).

[22] G. Declerck, R, van Overstraeten, and G. Broux, "Measurement of Low Densities of Surface States at the Si-SiO$_2$-Interface," Solid-State Electronics, 16, 1451 (1973).

[23] N. S. Saks and M. G. Ancona, "Determination of Interface Trap Capture Cross Sections Using Three-Level Charge Pumping," IEEE Electron Device Lett., 11, 339 (August, 1990).

SOURCE LISTING

(1) Review articles on a variety of R–G topics were published in Solid-State Electronics, 21, (1978). Relevant papers include:
   (a) N. F. Mott, "Recombination: A Survey," Solid-State Electronics, 21, 1275 (1978).
   (b) H. J. Queisser, "Recombination at Deep Traps," Solid-State Electronics, 21, 1495 (1978).

(2) Perhaps the most authoritative and complete treatment of single- and multi-level R–G center statistics can be found in:
   (a) C. T. Sah, "The Equivalent Circuit Model in Solid-State Electronics–Part I: The Single Energy Level Defect Centers," Proc. IEEE, 55, 654 (May, 1967).
   (b) C. T. Sah, "The Equivalent Circuit Model in Solid-State Electronics–Part II: The Multiple Energy Level Impurity Centers," Proc. IEEE, 55, 672 (May, 1967).

PROBLEMS

GENERAL NOTE: Problems 5.5, 5.6, 5.7, and 5.9 require a working knowledge of quasi-Fermi levels. Quasi-Fermi levels are reviewed herein in Subsection 6.3.1.

5.1 Answer the following questions as concisely as possible.
   (a) Using the energy band diagram, indicate how one visualizes photogeneration, intrinsic Auger recombination, and recombination via SRH centers.
   (b) Prior to processing, a portion of a semiconductor sample contains $N_D = 10^{14}/cm^3$ donors and $N_T = 10^{11}/cm^3$ R–G centers. After processing (say in the fabrication of a device), the same portion of the semiconductor contains $N_D = 10^{16}/cm^3$ donors and $N_T = 10^{10}/cm^3$ R–G centers. Did the processing increase or decrease the minority carrier lifetime? Explain.
   (c) Briefly explain the difference between "equilibrium" and "steady-state."
   (d) Make a plot of the net recombination rate (R) versus position inside the depletion region of a *pn*-junction diode maintained under equilibrium conditions.

5.2 A semiconductor contains bulk traps that introduce an R–G level at $E_T' = E_i$. Steady-state conditions prevail.
   (a) If c_n is roughly the same order of magnitude as c_p, confirm that:
      (i) $n_T \simeq N_T$ ... in an *n*-type semiconductor subject to low level injection.
      (ii) $p_T \simeq N_T$ ... in a *p*-type semiconductor subject to low level injection.
   (b) The semiconductor is uniformly illuminated such that $\Delta n = \Delta p \gg n_0$ or p_0.
      (i) Determine n_T/N_T if $c_n = c_p$.
      (ii) Determine n_T/N_T if $c_p \gg c_n$.

(c) The semiconductor is depleted $(n \to 0, p \to 0)$ and $e_n = e_p$. What fraction of the R–G centers will be filled for the specified situation?

5.3 Utilizing the general steady-state R-expression, confirm that band gap centers with $E_T{'}$ near E_i make the best R–G centers. [Assume, for example, that the semiconductor is n-type, low level injection conditions prevail, and $\tau_n = \tau_p = \tau =$ constant independent of the trap energy. Next consider how R would vary with $E_T{'}$ under the given conditions and conclude with a rough sketch of R versus $E_T{'}$ across the band gap.]

5.4 (a) Read and summarize the paper by W. Zimmerman, Electronics Letters, 9, 378 (August, 1973).

(b) Starting with the R-expression established in the text [Eq. (5.24)], derive Zimmerman's equation for τ.

5.5 The energy band diagram for a reverse-biased Si pn-junction diode under steady-state conditions is pictured in Fig. P5.5.

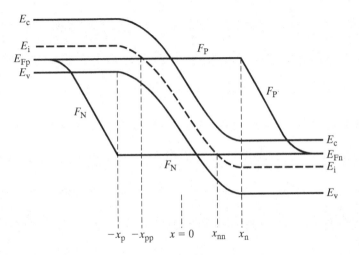

Figure P5.5

(a) With the aid of the diagram and assuming single-level R–G center statistics, $\tau_n = \tau_p = \tau$, and $E_T{'} = E_i$, simplify the general steady-state net recombination rate expression to obtain the simplest possible relationship for R at (i) $x = 0$, (ii) $x = -x_{pp}$, (iii) $x = x_{nn}$, (iv) $x = -x_p$, and (v) $x = x_n$.

(b) Sketch R versus x for x-values lying within the electrostatic depletion region $(-x_p \le x \le x_n)$.

(c) What was the purpose or point of this problem?

5.6 The energy band diagram for a forward-biased Si pn-junction diode maintained under steady-state conditions at room temperature $(T = 300 \text{ K})$ is pictured in Fig. P5.6. Note that $E_{Fn} - E_i = E_i - E_{Fp} = E_G/4$.

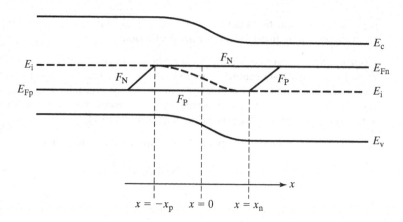

Figure P5.6

(a) For the particular situation pictured in Fig. P5.6, and assuming single-level R–G center statistics, $\tau_n = \tau_p = \tau$, and $E_T' = E_i$, show that the steady-state net recombination rate inside the electrostatic depletion region $(-x_p \le x \le x_n)$ can be simplified to

$$R \cong \frac{n_i}{\tau}\left[\frac{e^{E_G/4kT}}{e^{(F_N - E_i)/kT} + e^{(E_i - F_P)/kT}}\right]$$

(b) Plot $R/(n_i/\tau)$ versus x for $-x_p \le x \le x_n$. Assume the E_i variation between $x = -x_p$ and $x = x_n$ is approximately linear in constructing the plot.

(c) What is the purpose or point of this problem?

5.7 The energy band diagram for the semiconductor component of an MOS device under steady-state conditions is pictured in Fig. P5.7.

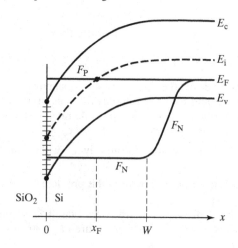

Figure P5.7

(a) Assuming single-level R–G centers, $\tau_n = \tau_p = \tau$, and $E_{T'} = E_i$, make a sketch of the net recombination rate (R) versus x for x-values lying within the electrostatic depletion region $(0 \leq x \leq W)$. Include the specific values of R at $x = 0$, $x = x_F$ and $x = W$ on your sketch.

(b) Is the net surface recombination rate (R_s) for the situation pictured in Fig. P5.7 expected to be less than, approximately equal to, or greater than the R_s at a totally depleted $(n_s \to 0,\ p_s \to 0)$ surface? Briefly explain how you arrived at your answer.

5.8 (a) For a totally depleted semiconductor surface $(n_s \to 0,\ p_s \to 0)$ with c_{ns}, c_{ps}, and $D_{IT}(E)$ all approximately constant over the midgap region, confirm the text assertion that

$$G_s \equiv -R_s = s_g n_i$$

where

$$s_g = (\pi/2)\sqrt{c_{ns}c_{ps}}\, kT D_{IT}$$

s_g is the surface generation velocity; all parameters in the s_g expression are evaluated at midgap.

(b) Based on the data presented in Fig. 5.13 and Fig. 5.14, estimate the expected value of s_g for an optimally processed, (100)-oriented, thermally oxidized Si surface.

5.9 The surface of an n-bulk solar cell is subjected to intense illumination giving rise to high level injection $(\Delta n_s = \Delta p_s \gg n_{s0})$. The surface region of the device is characterized by the energy band diagram shown in Fig. P5.9.

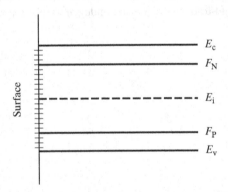

Figure P5.9

(a) What fraction of the surface centers at $E = F_N$, $E = E_i$, and $E = F_P$ will be filled for the specified steady-state situation? (Let D_{Filled} be the number of filled surface centers per eV-cm^2 at E. The fraction of filled surface centers is then D_{Filled}/D_{IT}.) Sketch D_{Filled}/D_{IT} versus E for $E_v \leq E \leq E_c$. Assume $c_{ns} = c_{ps}$ for all E in this part of the problem.

(b) Defining $R_s = s^* \Delta p_s$, obtain an *approximate closed-form* expression for s^* involving only $\Delta E \equiv F_N - F_P$ and the parameters c_{ns}, c_{ps}, and D_{IT} evaluated at midgap. Invoke and record all reasonable assumptions and simplifications needed to complete this problem.

5.10 In this problem we examine the statistics of thermal *band-to-band* recombination–generation.

(a) Using the energy band diagram, indicate the possible electronic transitions giving rise to (i) band-to-band recombination and (ii) band-to-band generation.

(b) Taking the semiconductor to be nondegenerate and paralleling the text derivation of Eqs. (5.14), establish the general relationship

$$r_b = c_b(np - n_i^2)$$

where r_b is the net band-to-band recombination rate and c_b is the band-to-band recombination coefficient (units of cm³/sec). (Note from the nature of the band-to-band processes that the net electron and hole recombination rates are always equal; $r_{electron} = r_{hole} = r_b$. It also follows, of course, that there is no special relationship for steady-state conditions.)

(c) Show that the general r_b relationship reduces to

$$r_b = \frac{\Delta n}{\tau_b}$$

$$\tau_b \equiv \frac{1}{c_b(n_0 + p_0)}$$

under low level injection conditions where $\Delta n = \Delta p$.

(d) Given $c_b \cong 5 \times 10^{-15}$ cm³/sec in Si at 300 K, and assuming $\tau \leq 1$ msec due to recombination at R–G centers, does one have to worry about carrier recombination via the band-to-band process in nondegenerately doped Si at room temperature? Explain.

CHAPTER 6

Carrier Transport

In this chapter we complete the task of constructing the knowledge and analytical base required in the operational modeling of semiconductor devices. Specifically, we provide a description of the carrier motion and currents inside a semiconductor resulting from applied fields and gradients. Whereas the recombination–generation analysis examined carrier action associated with *vertical* transitions on the energy band diagram, the carrier transport analysis to be presented treats carrier action which occurs *horizontally* on the energy band diagram. Drift and diffusion, the two major mechanisms giving rise to carrier transport within a semiconductor, are individually addressed in Sections 6.1 and 6.2, respectively. The presentation in each case is supplemented by a consideration of related topics of practical importance. In Section 6.3 we combine results from the drift, diffusion, and recombination–generation analyses into an overall mathematical description of carrier action inside a semiconductor. The "equations of state" thereby established constitute the basic set of equations that must be solved subject to imposed boundary conditions to obtain the system variables [$n(t)$, $p(t)$, currents, etc.] within a semiconductor under nonequilibrium conditions.

6.1 DRIFT

6.1.1 Definition-Visualization

Drift, by definition, *is charged-particle motion in response to an applied electric field.* On a microscopic scale the drifting motion within semiconductors can be described as follows: When an electric field ($\mathscr{E}$) is applied across a semiconductor as visualized in Fig. 6.1(a), the force acting on the carriers tends to accelerate the $+q$ charged holes in the direction of the electric field and the $-q$ charged electrons in the direction opposite to the electric field. The carrier acceleration is frequently interrupted, however, by scattering events—collisions between the carriers and ionized impurity atoms, thermally agitated lattice atoms, or other scattering centers. The result, pictured in Fig. 6.1(b), is a net carrier motion along the direction of the electric field, but in a disjointed fashion involving repeated periods of acceleration and subsequent decelerating collisions.

The microscopic drifting motion of a single carrier is obviously complex and quite tedious to analyze in any detail. Fortunately, measurable quantities of interest are

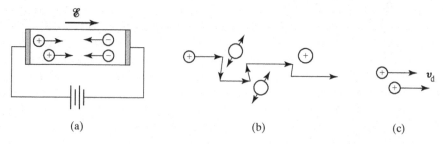

Figure 6.1   Visualization of carrier drift: (a) motion of carriers within a biased semiconductor bar; (b) drifting hole on a microscopic or atomic scale; (c) carrier drift on a macroscopic scale.

macroscopic observables which reflect the average or overall motion of the carriers. Averaging over all electrons or holes in a semiconductor bar at any given time, we find that the resultant motion of each carrier type can be described in terms of a drift velocity, v_d. In other words, on a macroscopic scale, drift can usually be visualized (see Fig. 6.1(c)) as nothing more than all carriers of a given type moving along at a constant velocity in a direction parallel or antiparallel to the applied electric field.

By way of clarification, it is important to point out that the drifting motion of the carriers arising in response to an applied electric field is actually superimposed upon the always-present thermal motion of the carriers. Being completely random, however, the thermal motion averages out to zero on a macroscopic scale, does not contribute to carrier transport, and can be conceptually neglected.

6.1.2 Drift Current

Let us next turn to the task of developing an analytical expression for the current flowing within a semiconductor as a result of carrier drift. By definition

I(current) = the charge per unit time crossing an arbitrarily chosen plane of
observation oriented normal to the direction of current flow.

Considering the p-type semiconductor bar of cross-sectional area A shown in Fig. 6.2, and specifically noting the arbitrarily chosen v_d-normal plane lying within the bar, we can argue:

$v_d t$	...	All holes this distance back from the v_d-normal plane will cross the plane in a time t;
$v_d t A$	...	All holes in this volume will cross the plane in a time t;
$p v_d t A$	...	Holes crossing the plane in a time t;
$q p v_d t A$	...	Charge crossing the plane in a time t;
$q p v_d A$	...	Charge crossing the plane per unit time.

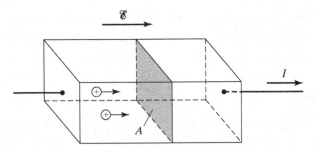

Figure 6.2 Expanded view of a biased *p*-type semiconductor bar of cross-sectional area *A*.

The word definition of the last quantity is clearly identical to the formal definition of current. Thus

$$I_{P|\text{drift}} = qp\,v_d\,A \tag{6.1}$$

or, in vector notation,

$$\mathbf{J}_{P|\text{drift}} = qp\,\boldsymbol{v}_d \tag{6.2}$$

where $\mathbf{J}$ is the *current density* and is equal in magnitude to the current per unit area ($J = I/A$).

Since the drift current arises in response to an applied electric field, it is reasonable to proceed one step further and seek a form of the current relationship which explicitly relates $\mathbf{J}_{P|\text{drift}}$ to the applied electric field. To this end we note that, for small to moderate values of $\mathscr{E}$, the measured drift velocity in semiconductors (see Fig. 6.3) is directly proportional to the applied electric field. Theoretical analyses of a more fundamental nature also arrive at the same conclusion. Excluding situations involving large $\mathscr{E}$-fields, we can therefore write

$$\boldsymbol{v}_d = \mu_p\mathscr{E} \tag{6.3}$$

where μ_p, the *hole mobility*, is the constant of proportionality between v_d and $\mathscr{E}$. Hence, substituting Eq. (6.3) into Eq. (6.2), one obtains

$$\boxed{\mathbf{J}_{P|\text{drift}} = q\mu_p p\mathscr{E}} \tag{6.4a}$$

By a similar argument applied to electrons, one finds

$$\boxed{\mathbf{J}_{N|\text{drift}} = q\mu_n n\mathscr{E}} \tag{6.4b}$$

where $\mathbf{J}_{N|\text{drift}}$ is the electron current density due to drift and μ_n is the electron mobility.

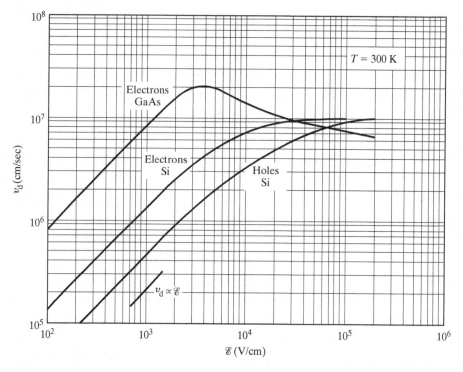

Figure 6.3   Measured drift velocity of carriers in high-purity Si and GaAs as a function of the applied electric field. Plots were constructed from 300 K data presented in ref. [1–3]

6.1.3 Mobility

As can be inferred from Eqs. (6.4), the electron and hole mobilities are central parameters in characterizing carrier transport due to drift. The carrier mobilities are in fact very important parameters which play a key role in characterizing the performance of many devices. It is reasonable therefore to examine these parameters in some detail to provide a core of useful information for future reference.

Basic Information

Standard Units: cm^2/V-sec.
Sample Numerical Values: $\mu_n \simeq 1360$ cm^2/V-sec and $\mu_p \simeq 460$ cm^2/V-sec at 300 K in $N_D = 10^{14}$/cm^3 and $N_A = 10^{14}$/cm^3 doped Si, respectively. In uncompensated high-purity (N_D or $N_A \leq 10^{15}$/cm^3) GaAs, the room-temperature drift mobilities are projected[3] to be $\mu_n \simeq 8000$ cm^2/V-sec and $\mu_p \simeq 320$ cm^2/V-sec. The quoted values are useful for comparison purposes and when performing order-of-magnitude computations. Also note that $\mu_n > \mu_p$ for both Si and GaAs. In the major semiconductors, μ_n is consistently greater than μ_p for a given doping and system temperature.

Physical Interpretation: Mobility is a measure of the ease of carrier motion within a semiconductor crystal. The lower the mobility of carriers within a given semiconductor, the greater the number of motion-impeding collisions.[†]

Theoretical Considerations

It should be obvious from the last basic-information item that the carrier mobility varies inversely with the amount of scattering taking place within the semiconductor. Simply stated, an increase in motion-impeding collisions leads to a decrease in mobility. To theoretically characterize the carrier mobility it is therefore necessary to consider the different types of scattering events that can take place inside a semiconductor. These include:

   (i) Phonon (lattice) scattering,

  (ii) Ionized impurity scattering,

 (iii) Scattering by neutral impurity atoms and defects,

 (iv) Carrier-carrier scattering, and

  (v) Piezoelectric scattering.

Of the scattering mechanisms cited, phonon and ionized impurity scattering tend to dominate in device-quality semiconductors. Phonon scattering refers to the collisions between the carriers and thermally agitated lattice atoms. The coulombic attraction or repulsion between the charged carriers and the ionized donors and/or acceptors leads to ionized impurity scattering. (Ionized deep-level centers can also give rise to ionized impurity scattering, but the concentration of the deep-level centers is typically negligible compared to the donor or acceptor concentration.)

The remaining scattering mechanisms are only important under certain conditions or enter indirectly into the overall scattering analysis. Neutral atom scattering, for example, becomes important at low temperatures where an appreciable fraction of the donors or acceptors are neutralized by carrier freeze-out. Defect scattering must be included when treating polycrystalline and other high-defect materials. Carrier-carrier scattering between electrons and holes is routinely insignificant because high concentrations of both carrier types is seldom present at the same point in a given semiconductor. Electron-electron and hole-hole scattering, on the other hand, do not affect the mobility directly because collisions between carriers of the same type cannot alter the total momentum of those carriers. However, same-carrier scattering randomizes the way the total momentum is distributed among the electrons or holes, and therefore has an indirect effect on other scattering mechanisms. This effect is normally handled as a modification to the expected phonon and ionized impurity scattering. Finally, piezoelectric

---

[†]Some care must be exercised in applying the noted physical interpretation when considering different semiconductors. It is readily established that $\mu = q\tau/m^*$, where τ is the mean free time between scattering events and m^* is the carrier effective mass. Thus carriers in a semiconductor with a smaller m^* will exhibit a higher mobility even though the number of motion-impeding collisions may be the same.

scattering is confined to piezoelectric materials such as GaAs where a displacement of the component atoms from their lattice sites gives rise to an internal electric field. GaAs, however, is only weakly piezoelectric and the associated scattering is relatively insignificant[4].

In analyses performed to compute the expected mobility from first principles it is common practice to associate a "component" mobility with each type of scattering process. For example, one introduces

$\mu_{Ln}(\mu_{Lp})$ ... the electron (hole) mobility that would be observed if only lattice scattering existed.

$\mu_{In}(\mu_{Ip})$ ... the electron (hole) mobility that would be observed if only ionized impurity scattering existed.

$\vdots$ $\vdots$

For the typically dominant phonon and ionized impurity scattering, single-component theories yield, respectively, to first order[5,6]

$$\mu_L \propto T^{-3/2} \tag{6.5}$$

and

$$\mu_I \propto T^{3/2}/N_I \tag{6.6}$$

where $N_I \equiv N_D^+ + N_A^-$.

It is worthwhile to comment that the general forms of the dependencies noted in Eqs. (6.5) and (6.6) are readily understood on an intuitive basis. Lattice vibrations, for one, would be expected to increase with temperature, thereby enhancing the probability of lattice scattering and reducing the associated mobility. Elevating the temperature, on the other hand, increases the thermal velocity of the carriers, which in turn reduces the time a carrier spends in the vicinity of an ionized impurity center. The less time spent in the vicinity of an ionized scattering center, the smaller the deflection, the smaller the effect of the scattering event, and the greater the expected value of μ_I. Increasing the number of ionized scattering centers, of course, proportionately increases the probability of scattering and decreases μ_I.

Once expressions for μ_{Ln}, μ_{In}, etc., have been established, the carrier mobility for a given doping and T is obtained by appropriately combining the component mobilities. Noting that each scattering mechanism gives rise to a "resistance-to-motion" which is inversely proportional to the component mobility, and taking the "resistances" to be simply additive (analogous to a series combination of resistors in an electrical circuit), one obtains

$$\frac{1}{\mu_n} = \frac{1}{\mu_{Ln}} + \frac{1}{\mu_{In}} + \cdots \tag{6.7a}$$

$$\frac{1}{\mu_p} = \frac{1}{\mu_{Lp}} + \frac{1}{\mu_{Ip}} + \cdots \tag{6.7b}$$

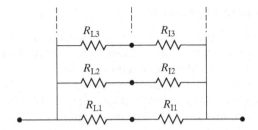

Figure 6.4   Electrical analog of the scattering "resistance" in semiconductors. The R_{Lj}'s and R_{Ij}'s model, respectively, the phonon and ionized-impurity scattering at different carrier energies.

The foregoing oft-quoted combination of component mobilities, sometimes referred to as Matthiessen's rule, is actually a first-order approximation. This is true because the resistances-to-motion vary with the thermal velocity (or energy) of the carriers, and the various scattering mechanisms exhibit a different velocity dependence. The situation can be understood with the aid of the analogous electrical circuit shown in Fig. 6.4. The R_{Lj}'s and R_{Ij}'s in this figure respectively model the phonon and ionized impurity scattering at different carrier energies. In applying Matthiessen's rule to compute the overall resistance, one is effectively tying all of the central nodes in Fig. 6.4 together, which is strictly valid only if $R_{Lj}/R_{Ij} = R_{L1}/R_{I1}$ (j = 2, 3, $\cdots$). Taking into account the cited energy dependence, and assuming phonon and impurity ion scattering to be dominant, one finds the total mobilities in germanium and silicon are more accurately approximated by the combinational relationship[7]

$$\mu = \mu_L\{1 + x^2[\text{Ci}(x)\cos x + (\text{Si}(x) - \pi/2)\sin x]\} \qquad (6.8)$$

where $x^2 = 6\mu_L/\mu_I$ and $\text{Ci}(x)$, $\text{Si}(x)$ are the cosine and sine integrals of x, respectively.

The situation for GaAs is even more complex. In GaAs, longitudinal optical phonons contribute significantly to the overall scattering. [Lattice vibrations are divisible into four sets of normal modes or types of phonons. Longitudinal and transverse lattice vibrations each give rise to a lower-energy (acoustical) and higher-energy (optical) set of modes.] For scattering by the longitudinal-optical mode, the resistances-to-motion at different carrier velocities are interdependent. The R_{Lj}'s in Fig. 6.4 may be viewed as being dependent on the voltage drop across the resistors. Accordingly, it is necessary to use an iterative or variational procedure to accurately combine all the relevant scattering mechanisms: there is no simple combinational formula.

The culmination of the theoretical analysis is of course the numerical computation of the expected carrier mobilities, calculations that incorporate all relevant scattering mechanisms and the proper combination of scattering components. Computational results of this type have been reported by Li et al.[8-10] for Si and by Walukiewicz et al.[4] for electrons in GaAs. Generally, the predicted results are in excellent agreement with experimental observations over a wide range of temperatures and dopings. For details and additional mobility/scattering information, the interested reader is referred to the cited literature.

Doping/Temperature Dependence

Knowledge of the doping and temperature dependence of the carrier mobilities is often an absolute necessity in the modeling and design of devices. The required doping and temperature information could be derived of course from theoretical computations or from experimental data available in the device literature. As we have noted, however, theoretical computations can become rather involved (and it is difficult to pick off accurate values of the mobility from those little logarithmic plots published in journals). Experimental data, on the other hand, is always prone to error, and there may be some question as to which set of conflicting data to utilize. For these reasons there have evolved surprisingly accurate "empirical-fit" relationships that are widely employed to deduce the expected carrier mobilities at a given doping and temperature. Valid over the range of normally encountered dopings and typical operating temperatures, the empirical-fit expressions are especially convenient when performing computer simulations. The form of a relationship is established on an empirical basis by noting general functional dependencies. Parameters in the relationship are next adjusted until one obtains an acceptable fit to the best available experimental data or first principle theoretical computation.

The Si carrier mobility versus doping and temperature plots presented respectively in Figs. 6.5 and 6.6 were constructed employing the empirical-fit relationship[11,12]

$$\mu = \mu_{\text{min}} + \frac{\mu_0}{1 + (N/N_{\text{ref}})^\alpha} \tag{6.9}$$

where μ is the carrier mobility, N is the doping (either N_A or N_D), and all other quantities are fit parameters that exhibit a temperature dependence of the form

$$A = A_0(T/300)^\eta \tag{6.10}$$

A_0 is a temperature-independent constant (the 300 K value of a parameter), T is the temperature in Kelvin, and η is the temperature exponent for the given fit parameter. Table 6.1 lists the A_0 and η values utilized in constructing Figs. 6.5 and 6.6.[†]

In examining Fig. 6.6, note that, for temperatures between $-50°C$ and $200°C$, there is a monotonic falloff in mobility with increasing N_A or N_D. The decrease in mobility with increasing dopant concentrations is of course caused by an increase in ionized impurity scattering. At the lower dopings the mobility approaches the doping-independent limiting value set by lattice scattering ($\mu_{\text{max}} = \mu_L$). The temperature behavior displayed in Fig. 6.6 is also readily understood. In lightly doped material, lattice scattering dominates and the Si carrier mobilities decrease with temperature roughly as $T(K)^{-\eta}$. This is clearly evident from the inserts in Fig. 6.6. Note, however,

---

[†]The room-temperature hole parameters and the temperature exponents listed in Table 6.1 were taken from ref. [12]. The electron parameter prefactors given in ref. [12], however, yield room-temperature mobilities that are lower than those observed experimentally and predicted theoretically, especially at low dopings. For this reason we have chosen to employ the room-temperature electron parameters quoted in ref. [13].

Table 6.1   Mobility Fit Parameters for Si[12,13]

| Parameter | T-independent Prefactor | | Temperature Exponent |
	Electrons	Holes	
N_{ref}(cm^{-3})	1.30×10^{17}	2.35×10^{17}	2.4
μ_{min}(cm^2/V-s)	92	54.3	−0.57
μ_0(cm^2/V-s)	1268	406.9	−2.33 electrons −2.23 holes
α	0.91	0.88	−0.146

that the Fig. 6.6 inserts reflect the fact that experiments and more exacting theories give $\mu_L \propto T^{-2.3 \pm 0.1}$ for electrons and $\mu_L \propto T^{-2.2 \pm 0.1}$ for holes rather than the $T^{-3/2}$ expected from a first-order analysis. With increased doping, impurity-ion scattering becomes more and more important. Whereas lattice scattering increases with increasing T, impurity-ion scattering decreases. In heavily doped material, these opposing temperature dependencies partially cancel, yielding a mobility that is far less temperature-sensitive.

Finally, the theoretically predicted electron mobility in GaAs as a function of the electron concentration (equal to the net doping concentration) for selected compensation ratios ($\theta = N_A/N_D$) is graphed in Fig. 6.7. It should be explained that some degree of unintentional dopant compensation is a likely occurrence in GaAs. Figure 6.7 was constructed utilizing the first principle results of ref. [4] which were presented in a convenient tabular form. It is possible, however, to obtain a close fit to the computational results using Eq. (6.9) and the fit parameters listed in Table 6.2[14].

6.1.4 High-Field/Narrow-Dimension Effects

Fields of 10^4 to 10^5 V/cm are readily attained within the depletion region of pn-junctions and in the pinch-off region of field-effect transistors. As is evident from Fig. 6.3, when the $\mathscr{E}$- field inside a semiconductor exceeds $1 - 5 \times 10^3$ V/cm, the drift velocity is no longer directly proportional to the applied electric field, and the mobility-based formulation begins to break down. Moreover, with the push to achieve higher operating speeds and increased packing densities, devices are being fabricated that have submicron ($<10^{-4}$ cm) lateral dimensions. The mobility-based formulation, and even the drift velocity concept itself, can fail when one attempts to describe carrier transport across narrow spatial dimensions. In this subsection we briefly examine drift-related phenomenological effects of practical importance that come into play in the modeling of carrier transport under high-field conditions or across narrow spatial dimensions.

Velocity Saturation

When the electric field inside a semiconductor is made progressively larger and larger, the drift velocity of the carriers tends to saturate or approach a field-independent constant value. Although not directly analogous, the situation may be likened to an object falling through the atmosphere toward the surface of the earth. Because of the frictional force exerted by the air, all free-falling objects eventually attain a terminal or maximum velocity. In semiconductors the limiting high-field drift velocity is referred to as the *saturation velocity*, v_{dsat}.

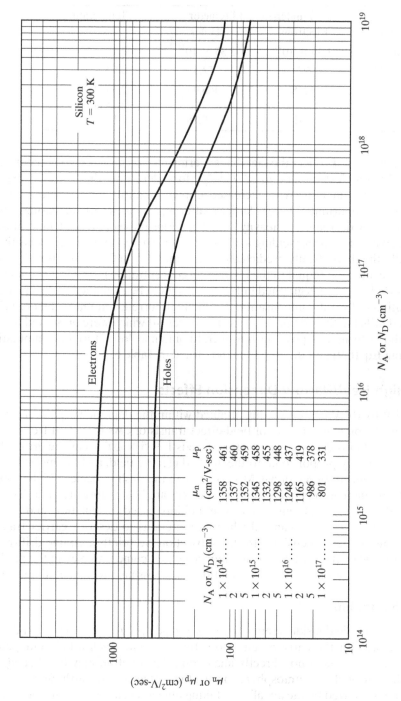

Figure 6.5 Room-temperature carrier mobilities in silicon as a function of the dopant concentration. μ_n is the electron mobility; μ_p is the hole mobility. [Mobility values were computed using Eq. (6.9) and the fit parameters listed in Table 6.1.]

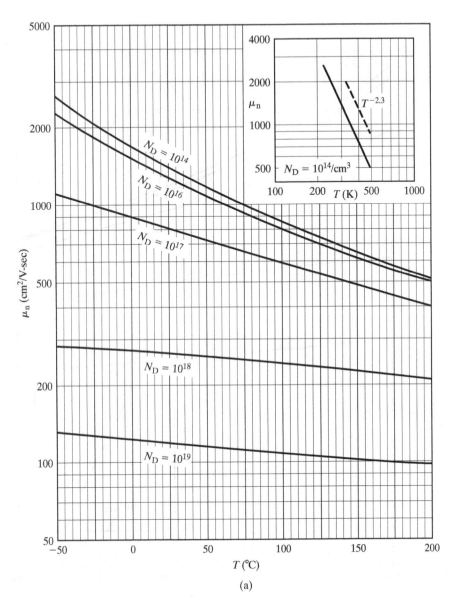

Figure 6.6 Temperature dependence of (a) electron and (b) hole mobilities in silicon for sample dopings ranging from $10^{14}/cm^3$ to $10^{19}/cm^3$. In the lightest-doped sample, $\mu \rightarrow \mu_L \propto T(K)^{-\eta}$. This is clearly evident from the insert plots of log μ versus log $T(K)$. [Mobility values were computed using Eq. (6.9) and the fit parameters listed in Table 6.1.]

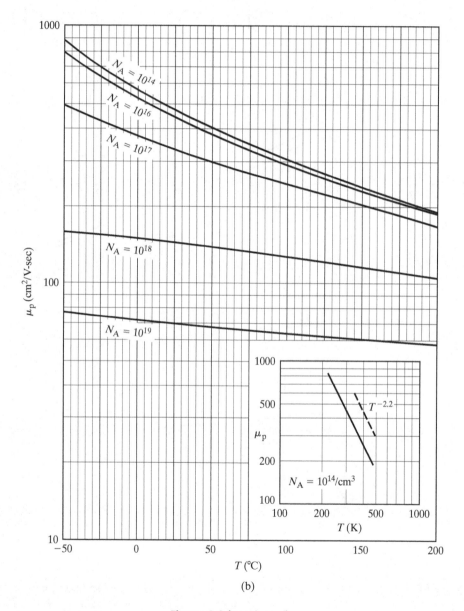

Figure 6.6 (continued)

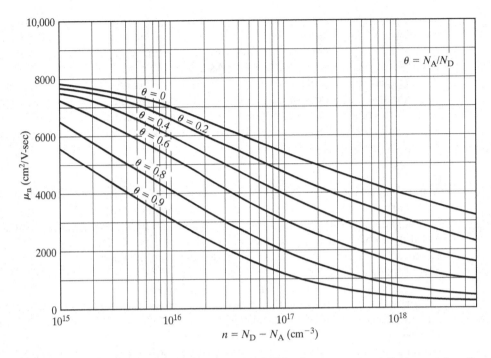

Figure 6.7 Theoretically predicted electron mobilities in compensated GaAs as a function of the net doping concentration. $\theta = N_A/N_D$ is the compensation ratio. (Data from Walukiewicz et al.[4])

Within Si at 300 K (see Fig. 6.3), $v_{dsat} \simeq 10^7$ cm/sec for both electrons and holes and occurs at an $\mathscr{E}$-field of approximately 10^5 V/cm. The observed temperature dependence of v_{dsat} for electrons in Si can be modeled by the empirical-fit expression[1]

$$v_{dsat} = \frac{v_{dsat}^0}{1 + Ae^{T/T_d}} \tag{6.11}$$

where $v_{dsat}^0 = 2.4 \times 10^7$ cm/sec, $A = 0.8$, $T_d = 600$ K, and T is the temperature in Kelvin. In GaAs the electron drift velocity actually exhibits a retrograde behavior,

Table 6.2 Mobility Fit Parameters for Electrons in GaAs[14] ($T = 300$ K)

$\theta = N_A/N_D$	μ_{min} (cm²/V-sec)	μ_0 (cm²/V-sec)†	α	N_{ref} (cm⁻³)
0.0	2750	5450	0.553	9.85×10^{16}
0.2	1750	6450	0.537	8.10×10^{16}
0.4	1100	7100	0.542	5.09×10^{16}
0.6	550	7650	0.537	2.79×10^{16}
0.8	200	8000	0.551	9.85×10^{15}
0.9	100	8100	0.594	4.02×10^{15}

†$\mu_0 = \mu_{max} - \mu_{min}$; $\mu_{max} = 8200$ cm²/V-sec.

decreasing with increasing electric fields above ~3.3 × 10³ V/cm at 300 K. Moreover, complete saturation has not been observed even at the highest reported measurement field of 2.2 × 10⁵ V/cm. The last fact notwithstanding, an effective v_{dsat} ~ 10⁷ cm/sec is sometimes used in the modeling of GaAs devices.

v_{dsat} is typically encountered in analyses treating carrier transport through "depleted" regions of a device structure. Although the mobility-dependent current expressions [Eqs. (6.4)] are no longer valid under high-field conditions, Eq. (6.2) and its electron analog are still correct. Thus, in a depletion region where $\mathscr{E}$ ~ 10⁵ V/cm, one can write

$$\mathbf{J}_{P|drift} = qp\mathbf{v}_{dsat} \qquad (6.12)$$

In field-effect transistors biased into the pinch-off regime, it is usually assumed that the drain current is limited by the channel conductivity. However, under certain conditions normally encountered in short-channel devices, the drain current can be limited by the amount of charge transported across the depleted pinch-off region. Eq. (6.12) or its electron analog then comes into play in determining the current observed at the terminals of the device.

v_{dsat} also comes into play in analyzing the maximum frequency response of bipolar junction transistors and other *pn*-junction devices. One factor that can limit the frequency response of a PIN photodiode, for example, is the amount of time it takes for photogenerated carriers to drift across the high-field "I" region. If the field in the I-region is sufficiently high, the transit time is of course just equal to the width of the I-region divided by v_{dsat}.

Intervalley Carrier Transfer

The reader has undoubtedly noticed the peaked nature of the drift velocity versus $\mathscr{E}$-field curve for electrons in GaAs. This interesting feature is a direct manifestation of intervalley electron transfer. As described in Chapter 3, the GaAs conduction band minimum occurs at the Γ-point, at $k = 0$. It was also pointed out that secondary minima exist at the Brillouin zone boundary along ⟨111⟩ directions. Lying 0.29 eV above the Γ-valley, these L-valleys are sparsely populated under equilibrium conditions at room temperature. When an accelerating field is applied to the crystal, however, the Γ-valley electrons gain energy between scattering events. If the maximum energy gain of an electron is in excess of 0.29 eV, intervalley transfer becomes possible and the population of the L-valley is enhanced at the expense of the Γ-valley (see Fig. 6.8). It is important to note that the effective mass of electrons at the center of the Γ-valley is $0.0632\,m_0$. The L-valleys, on the other hand, are ellipsoids with longitudinal and transverse effective masses of $m_\ell^* \simeq 1.9m_0$ and $m_t^* \simeq 0.075m_0$, respectively[3]. For drift-related considerations, an isotropic effective mass of $m^* = 0.55m_0$ may be assigned to the L-valley electrons. In other words, the effective mass of an electron increases by about an order of magnitude, and its drift velocity correspondingly decreases, upon transferring from the Γ-valley to the L-valley. At an applied field of approximately $\mathscr{E}_c = 3.3 × 10^3$ V/cm and $T = 300$ K, the decrease in drift velocity associated with the transfer of electrons to the L-valleys is sufficient to balance the routine increase in drift velocity resulting from an incremental increase in $\mathscr{E}$—i.e., under the specified conditions, v_d attains its peak value of ~2 × 10⁷ cm/sec. Increasing $\mathscr{E}$ above $\mathscr{E}_c$ further enhances

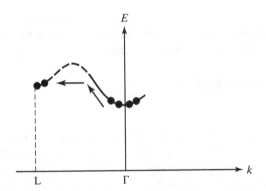

Figure 6.8 Intervalley electron transfer in GaAs.

the population of the L-valleys, which in turn gives rise to the observed decrease in drift velocity for electric fields in excess of $\mathscr{E}_c$.

The retrograde v_d versus $\mathscr{E}$ behavior or negative differential mobility exhibited by GaAs and a limited number of other materials (notably InP) provides the phenomenological basis for an important class of microwave devices, the Transferred-Electron Devices (TED's). TED's have been extensively used as local oscillators and power amplifiers at microwave frequencies ranging from 1 to 200 GHz.

Ballistic Transport/Velocity Overshoot

In visualizing carrier drift we implicitly assumed that the carriers experienced numerous scattering events between the point of injection into the semiconductor and the point of extraction from the semiconductor. This is equivalent to assuming the total distance (L) through which the carriers travel is much much greater than the mean distance (l) between scattering events. One's ability to define an average drift velocity, the entire mobility/drift-velocity formalism for that matter, begins to break down when $L \sim l$. For one, a phenomenon called *velocity overshoot* occurs, and the average carrier velocity can be substantially greater than that naively expected from a v_d versus $\mathscr{E}$ plot. Furthermore, in structures where $L < l$, a significant percentage of the carriers could conceivably travel from the point of injection to the point of extraction without experiencing a single scattering event. The carriers then would behave like projectiles similar to the electrons in a vacuum tube. The motion of the carriers under such conditions is referred to as *ballistic transport*. Ballistic transport is of interest because it would lead to super-fast devices.

At room temperature, deviations from the mobility/drift-velocity formalism are theoretically predicted in GaAs structures where $L \lesssim 1\ \mu$m and in Si structures where $L \lesssim 0.1\ \mu$m. As of this writing, commercial Si field effect transistors are being fabricated with channel lengths $\cong 0.1\ \mu$m, while transistors fabricated in research laboratories have been reported with channel lengths as small as 15 nm = 0.015 μm. Velocity overshoot does indeed occur in these structures. However, there is still a considerable amount of scattering. This takes place because the probability of scattering increases as the carrier velocity increases. Even a small bias applied across the channel length gives rise to a large increase in the scattering rate and a significantly reduced l. Ballistic transport, nevertheless, remains a topic of interest as device dimensions and operating voltages continue to decrease.

6.1.5   Related Topics

Resistivity

Resistivity is an important material parameter that is closely related to carrier drift. Qualitatively, resistivity is a measure of a material's inherent resistance to current flow—a "normalized" resistance that does not depend on the physical dimensions of the material. Quantitatively, resistivity (ρ) is defined as the proportionality constant between the electric field impressed across a homogeneous material and the total particle current per unit area flowing in the material; that is,

$$\mathscr{E} = \rho \mathbf{J} \tag{6.13}$$

or

$$\mathbf{J} = \sigma \mathscr{E} = \frac{1}{\rho} \mathscr{E} \tag{6.14}$$

where $\sigma = 1/\rho$ is the material conductivity. In a homogeneous material, $\mathbf{J} = \mathbf{J}_{\text{drift}}$ and, as established with the aid of Eqs. (6.4),

$$\mathbf{J}_{\text{drift}} = \mathbf{J}_{\text{N}|\text{drift}} + \mathbf{J}_{\text{P}|\text{drift}} = q(\mu_n n + \mu_p p)\mathscr{E} \tag{6.15}$$

It therefore follows that

$$\boxed{\rho = \frac{1}{q(\mu_n n + \mu_p p)}} \tag{6.16}$$

In a nondegenerate donor-doped semiconductor maintained in the extrinsic temperature region where $N_D \gg n_i$, $n \simeq N_D$, and $p \simeq n_i^2/N_D \ll n$. This result was established in Subsection 4.5.2. Thus, for typical dopings and mobilities, $\mu_n n + \mu_p p \simeq \mu_n N_D$ in an n-type semiconductor. Similar arguments yield $\mu_n n + \mu_p p \simeq \mu_p N_A$ in an $N_A \gg N_D$ p-type semiconductor. Consequently, under conditions normally encountered in Si samples maintained at or near room temperature, Eq. (6.16) simplifies to

$$\boxed{\rho = \frac{1}{q\mu_n N_D}} \qquad \ldots n\text{-type semiconductor} \tag{6.17a}$$

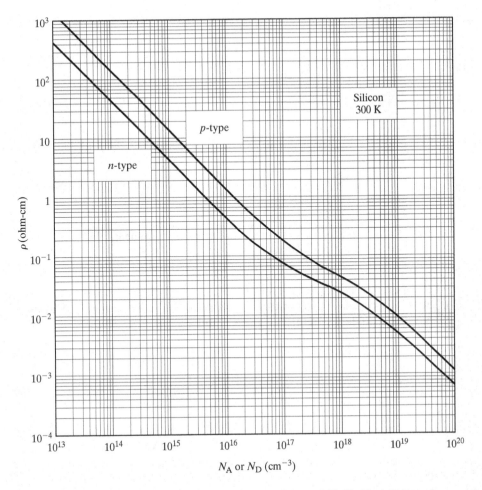

Figure 6.9 Si resistivity versus impurity concentration at 300 K. Resistivity values were computed employing Eqs. (6.9) and (6.17). The pictured curves are essentially identical to those presented in references [8], [10], [15] and [16].

and

$$\left[\rho = \frac{1}{q\mu_p N_A} \right] \quad \ldots p\text{-type semiconductor} \qquad (6.17b)$$

When combined with mobility-versus-doping data, Eqs. (6.17) provide a one-to-one correspondence between the resistivity, a directly measurable quantity, and the doping inside a semiconductor. In conjunction with plots of ρ versus doping (see Fig. 6.9), the measured resistivity is in fact routinely used to determine N_A or N_D in silicon samples.

A widely employed method for measuring semiconductor resistivities is the four-point probe technique. This technique is easy to implement, nondestructive, and especially convenient for probing the wafers used in device fabrication. In the standard

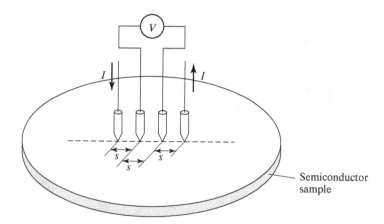

Figure 6.10   Schematic drawing of the probe arrangement, placement, and biasing in the standard four-point probe measurement.

four-point probe technique, four collinear, evenly spaced probes, as shown in Fig. 6.10, are brought into contact with the surface of the semiconductor. A known current I is passed through the outer two probes and the potential V thereby developed is measured across the inner two probes. The semiconductor resistivity is then computed from

$$\rho = 2\pi s \frac{V}{I}\mathscr{F}_c \tag{6.18}$$

where s is the probe-to-probe spacing and $\mathscr{F}_c$ is a well-documented "correction" factor. The correction factor typically depends on the thickness of the sample and on whether the bottom of the semiconductor is touching an insulator or a metal. For a full tabulation of correction factors the interested reader is referred to Chapter 4 in ref. [17]. We should note that semi-automatic instruments are available that compute the correction factor and display the resistivity after one inputs the sample thickness and measurement configuration.

Hall Effect

When there is a severe contact problem, the four-point probe technique cannot be used to measure the semiconductor resistivity. Moreover, additional information may be required to completely specify the concentration of dopants. For example, the mobility and hence the resistivity of GaAs typically depends not only on the net doping concentration but also on the degree of dopant compensation. For these reasons it is commonplace to combine resistivity measurements, involving metallurgical contacts, and Hall-effect measurements to determine the drift/doping parameters in GaAs and similarly constituted materials.

As envisioned in Fig. 6.11(a), Hall-effect measurements involve the application of a magnetic field (B) perpendicular to the direction of current flow in a semiconductor sample. The Hall effect itself is the appearance of a voltage drop (V_H) between the

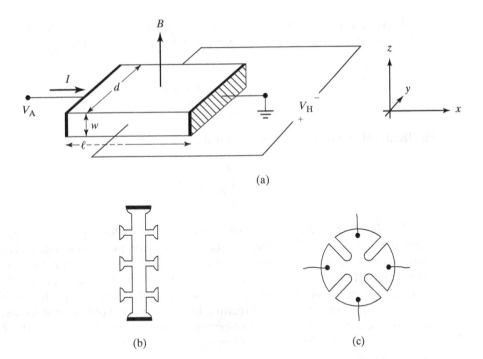

(a)

(b)   (c)

Figure 6.11 The Hall-effect measurement. (a) Measurement configuration assumed in discussing and analyzing the Hall effect. (b, c) Widely employed sample configurations: (b) the Hall-bar with "ears," and (c) a van der Pauw configuration.

faces of the sample normal to both the direction of current flow and the applied B-field. In discussing the Hall effect we will take the sample to be bar-like with x, y, and z dimensions of l, d, and w, respectively. V_A is the voltage applied across the ohmic contacts at the x-ends of the sample. The coordinate axes are assumed to be oriented such that current flows in the x-direction ($J_x = J = I/wd$), the magnetic field lies in the z-direction ($B_z = B$), and the Hall voltage or $\mathscr{E}$-field is developed along the y-direction ($\mathscr{E}_y = V_H/d$).

   Although we will provide a simple explanation for the appearance of the Hall voltage and will establish its relationship to fundamental material parameters, it is convenient to first introduce the Hall parameters that are normally computed from the raw measurement data. Specifically, the Hall coefficient (R_H), the resistivity (ρ) when $B = 0$, and the Hall mobility (μ_H) are routinely used to characterize experimental results. The Hall coefficient is calculated from the defining equation,

$$R_H \equiv \frac{\mathscr{E}_y}{J_x B_z} = \frac{V_H/d}{(I/wd)B} = \frac{V_H w}{BI} \tag{6.19a}$$

or

$$\boxed{R_H = \frac{10^8 V_H w}{BI}} \tag{6.19b}$$

if V_H is given in volts, w in cm, B in gauss, I in amps, and R_H in cm³/coul. With the B-field removed, the resistance of the bar is just $V_A/I = \rho l/wd$. Thus

$$\boxed{\rho = \frac{V_A}{I}\frac{wd}{l}} \tag{6.20}$$

Finally, the Hall mobility is computed from

$$\boxed{\mu_H \equiv \frac{|R_H|}{\rho}} \tag{6.21}$$

Physically, the Hall voltage arises because of the deflecting force ($\pm q\boldsymbol{v}_d \times \mathbf{B}$) associated with the applied B-field. For the configuration pictured in Fig. 6.11(a) and an assumed p-type sample, holes moving in the $+x$ direction with drift velocity $\boldsymbol{v}_d$ are initially deflected in the $-y$ direction. This deflection in turn causes a pileup of holes on the front face of the sample, a deficit of holes along the back face of the sample, and a growing electric field in the $+y$ direction. In a short period of time, a steady-state condition is reached where the pileup of carriers ceases and the force due to the $\mathscr{E}$-field in the y-direction just balances the deflecting force associated with the B-field applied in the z-direction.

Based on the foregoing simple model, and still assuming a p-type sample, one can write

$$\mathbf{F}_y = q\boldsymbol{v}_d \times \mathbf{B} + q\mathscr{E}_y = 0 \tag{6.22}$$

or

$$-v_d B_z + \mathscr{E}_y = 0 \tag{6.23}$$

where $\mathbf{F}_y$ is the total force exerted on the carriers in the y-direction under steady-state conditions. Moreover, making use of Eq. (6.2),

$$J_x = qpv_d \tag{6.24}$$

or

$$v_d = J_x/qp \tag{6.25}$$

Combining Eqs. (6.23) and (6.25) then yields

$$-\frac{J_x B_z}{qp} + \mathscr{E}_y = 0 \tag{6.26}$$

from which one infers

$$R_H \equiv \frac{\mathscr{E}_y}{J_x B_z} = \frac{1}{qp} \qquad \ldots \text{if } p \gg n \qquad (6.27a)$$

Similarly, one concludes that

$$R_H = -\frac{1}{qn} \qquad \ldots \text{if } n \gg p \qquad (6.27b)$$

Equations (6.27), relating the Hall coefficient computed from experimental data and the carrier concentrations, are understandably first-order results. Performing a more exacting analysis, one finds[17]

$$R_H = \begin{cases} \dfrac{r_H}{qp} & \ldots \text{if } p \gg n \qquad (6.28a) \\[2em] -\dfrac{r_H}{qn} & \ldots \text{if } n \gg p \qquad (6.28b) \end{cases}$$

where r_H is the *Hall factor*. The dimensionless Hall factor is in principle a theoretically computable quantity that varies with the material under analysis, the carrier type, the dominant scattering mechanisms, temperature, and the magnitude of the B-field. r_H typically assumes a value close to unity, is seldom less than 0.5 or greater than 1.5, and approaches unity as $B \to \infty$. In other words, the error introduced is not too great even if one has no *a priori* knowledge of r_H and arbitrarily sets $r_H = 1$.

With some manipulation, the relationships we have presented allow one to determine the desired material parameters. Consider, for example, an n-type GaAs sample maintained at room temperature. A negative Hall coefficient would of course confirm the n-type nature of the sample. Substituting the experimentally determined R_H into Eq. (6.28b) and solving for n next yields the electron concentration and hence the net doping concentration ($n \simeq N_D - N_A$). Since $\rho \simeq 1/q\mu_n n$ in an n-type sample, the ρ computed from the experimental data using Eq. (6.20) and the n deduced from the Hall coefficient are all one requires to determine μ_n. Finally, knowing μ_n and $N_D - N_A$, one can use the theoretical curves presented in Fig. 6.7 to estimate the degree of dopant compensation.

As a practical matter, the reader should be cautioned that the majority of electron and hole mobilities quoted in the GaAs literature are actually Hall mobilities. Per Eq. (6.21), the Hall mobility can be determined directly from experimental data without a prior knowledge of the Hall factor. The relationship between the Hall mobility and the drift mobility (μ_n or μ_p) is established by substituting $\rho = 1/q\mu_n n$ or $\rho = 1/q\mu_p p$ and the appropriate Eq. (6.28) into Eq. (6.21). One finds

$$\mu_H = r_H \mu_{\text{drift}} \qquad (6.29)$$

Since $r_H \geq 1$ for both electrons and holes in GaAs, the drift mobilities are invariably smaller than the corresponding Hall mobilities.

A comment is also in order concerning the sample configurations employed in Hall-effect measurements. One extensively used configuration, a bar with protruding "ears," is pictured in Fig. 6.11(b). The ears facilitate contact attachment and provide measurement redundancy. A configuration like the one shown in Fig. 6.11(c), however, provides a very popular alternative for routine material characterization. This configuration takes advantage of the van der Pauw method[17,18] for measuring the resistivity of flat, arbitrarily shaped samples. Other than electrical data, the van der Puw method requires only four ohmic contacts along the periphery of the sample and knowledge of the sample's thickness.

6.2   DIFFUSION

6.2.1   Definition-Visualization

Diffusion is a process whereby particles tend to spread out or redistribute as a result of their random thermal motion, migrating on a macroscopic scale from regions of high particle concentration into regions of low particle concentration. If allowed to proceed unabated, the diffusion process operates so as to produce a uniform distribution of particles. The diffusing entity, it should be noted, need not be charged: thermal motion, not interparticle repulsion, is the enabling action behind the diffusion process.

Figure 6.12(a) provides an idealized visualization of diffusion on a microscopic scale. For simplicity, the system under investigation has been taken to be one-dimensional. As pictured in Fig. 6.12(a), the randomness of the thermal motion gives rise to an equal

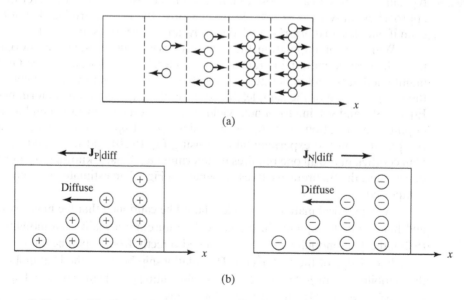

Figure 6.12   (a) Idealized visualization of particle diffusion on a microscopic scale. (b) Visualization of electron and hole diffusion on a macroscopic scale.

number of particles moving in the $+x$ and $-x$ directions within any given Δx-section of the system. Thus, there will be an equal *outflow* of particles per second from any interior section into neighboring sections on the right and left. Because of the concentration gradient, however, the number of particles per second *entering* an interior section from the right will be greater than the number per second entering from the left. It logically follows that over a period of time (and assuming the particle motions are randomized by collisions after entering a new Δx-section), the particle concentrations in Δx-sections on the left will progressively increase at the expense of the particle concentrations in the Δx-sections on the right.

On a macroscopic scale one merely observes the overall migration of particles from regions of high particle concentration to regions of low particle concentration. Within semiconductors the mobile particles, the electrons and holes, are charged, and diffusion-related carrier transport therefore gives rise to particle currents as pictured in Fig. 6.12(b).

6.2.2   Diffusion Current

Although conceptually straightforward, the derivation of analytical expressions for the carrier diffusion currents can become rather involved mathematically. An exacting derivation takes into account, for example, the distribution of carrier velocities within a semiconductor and the three-dimensional character of the thermal motion. Herein we present a highly simplified derivation which nevertheless retains the critical features of a more exacting analysis. A more precise treatment of particle diffusion can be found in the book by Present[19].

SIMPLIFYING ASSUMPTIONS:

In the derivation to be presented we make the following simplifications: (1) Carrier motion and concentration gradients are restricted to one-dimension (the x-direction). Because the motion is random and the system one-dimensional, half of the carriers move in the $+x$ direction and half move in the $-x$ direction. (2) All carriers move with the same velocity, $\bar{v}$. In reality there is a distribution of carriers within a semiconductor; we are conceptually placing all of the carriers at the distribution mean—$\bar{v} = v_{th}$, where v_{th} is the average thermal velocity introduced in Chapter 5. (3) The distance traveled by carriers between collisions (scattering events) is a fixed length, l. Moreover, all carriers are assumed to scatter at the same instant—that is, all carriers scatter in unison, move a distance l, scatter in unison, move a distance l, etc. If correlated to the true situation inside a semiconductor, l corresponds to the *mean* distance traveled by the carriers between scattering events.

DERIVATION PROPER:

Consider the p-type semiconductor bar of cross-sectional area A and the steady-state hole concentration gradient shown respectively in Fig. 6.13(a) and (b). Under the simplifying assumptions we have made, if t is arbitrarily set equal to zero at the instant all of the carriers scatter, it follows that half of the holes in a volume lA on either side

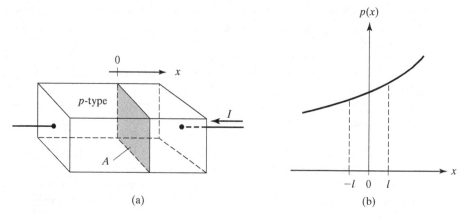

Figure 6.13   (a) p-type semiconductor bar of cross-sectional area A and (b) steady-state hole concentration about $x = 0$ envisioned in deriving the hole current associated with diffusion.

of $x = 0$ will be moving in the proper direction so as to cross the $x = 0$ plane prior to the next scattering event at $t = l/\bar{v}$. Thus we can write

$$\vec{p} = \begin{pmatrix} \text{Holes moving in the } +x \\ \text{direction which cross the} \\ x = 0 \text{ plane in a time } l/\bar{v} \end{pmatrix} = \frac{A}{2} \int_{-l}^{0} p(x)\mathrm{d}x \qquad (6.30\text{a})$$

$$\overleftarrow{p} = \begin{pmatrix} \text{Holes moving in the } -x \\ \text{direction which cross the} \\ x = 0 \text{ plane in a time } l/\bar{v} \end{pmatrix} = \frac{A}{2} \int_{0}^{l} p(x)\mathrm{d}x \qquad (6.30\text{b})$$

Since l is typically quite small, the first two terms in a Taylor series expansion of $p(x)$ about $x = 0$ will closely approximate $p(x)$ for x values between $-l$ and $+l$; that is,

$$p(x) \simeq p(0) + \frac{\mathrm{d}p}{\mathrm{d}x}\bigg|_{0} x \qquad \dots -l \le x \le l \qquad (6.31)$$

Substituting Eq. (6.31) into Eqs. (6.30) and performing the integrations yields

$$\vec{p} = \frac{1}{2}Alp(0) - \frac{1}{2}A\frac{\mathrm{d}p}{\mathrm{d}x}\bigg|_{0}\frac{l^2}{2} \qquad (6.32\text{a})$$

$$\overleftarrow{p} = \frac{1}{2}Alp(0) + \frac{1}{2}A\frac{\mathrm{d}p}{\mathrm{d}x}\bigg|_{0}\frac{l^2}{2} \qquad (6.32\text{b})$$

and

$$\vec{p} - \overleftarrow{p} = -A\frac{dp}{dx}\bigg|_0 \frac{l^2}{2} \tag{6.33}$$

Note that $\vec{p} - \overleftarrow{p}$ is the net number of $+ x$ directed holes that cross the $x = 0$ plane in a time $l/\bar{v}$. If $\vec{p} - \overleftarrow{p}$ is multiplied by q and divided by $t = l/\bar{v}$, one obtains the net charge crossing the $x = 0$ plane per unit time due to diffusion. In other words,

$$I_{P|\text{diff}} = \frac{q(\vec{p} - \overleftarrow{p})}{l/\bar{v}} \tag{6.34a}$$

$$= -\frac{1}{2}qA\bar{v}l\frac{dp}{dx} \tag{6.34b}$$

and

$$J_{P|\text{diff}} = -q(\bar{v}l/2)\frac{dp}{dx} \tag{6.35}$$

In writing down Eqs. (6.34b) and (6.35) we have dropped the "$|_0$" designation which indicated that dp/dx was to be evaluated at $x = 0$. This is acceptable since the $x = 0$ plane was merely chosen for convenience, and the same result is obtained for any $x = $ constant plane inside the semiconductor bar. Finally, introducing $D_P \equiv \bar{v}l/2$, we obtain

$$J_{P|\text{diff}} = -qD_P\frac{dp}{dx} \tag{6.36}$$

It is interesting to note that an exacting three-dimensional analysis leads to precisely the same x-component result except $D_P \equiv \bar{v}l/3$.

CONCLUSION:

Generalizing Eq. (6.36) to include a three-dimensional concentration gradient, one obtains

$$\boxed{\mathbf{J}_{P|\text{diff}} = -qD_P\nabla p} \tag{6.37a}$$

Analogously

$$\boxed{\mathbf{J}_{N|\text{diff}} = qD_N\nabla n} \tag{6.37b}$$

D_P and D_N are, respectively, the hole and electron *diffusion coefficients* with standard units of cm^2/sec.

6.2.3   Einstein Relationship

The diffusion coefficients are obviously central parameters in characterizing carrier transport due to diffusion. Given the importance of the diffusion coefficients, one might anticipate an extended examination of relevant properties paralleling the mobility presentation in Subsection 6.1.3. Fortunately, an extended examination is unnecessary because the D's and the μ's are interrelated. It is only necessary to establish the connecting formula known as the Einstein relationship.

Before performing the mathematical manipulations leading to the Einstein relationship, let us examine a *nonuniformly* doped semiconductor maintained under *equilibrium conditions*. A concrete example of what we have in mind is shown in Fig. 6.14. Figure 6.14(a) exhibits the assumed doping variation with position and Fig. 6.14(b) the corresponding equilibrium energy band diagram. Two facts inherent in the diagram are absolutely essential in establishing the Einstein relationship. (These facts are also important in themselves.) First of all,

> under equilibrium conditions the Fermi level inside a material (or inside a group of materials in intimate contact) is invariant as a function of position; that is $dE_F/dx = dE_F/dy = dE_F/dz = 0$ under equilibrium conditions.

As shown in Fig. 6.14(b), E_F is at the same energy for all x. The position independence of the Fermi energy is established by examining the transfer of carriers between allowed states with the same energy but at adjacent positions in an energy band. It is concluded the probability of filling the states at a given energy, $f(E)$, must be the same everywhere in the sample under equilibrium conditions. If this were not the case, carriers would preferentially transfer between states and thereby give rise to a net current. The existence of a net current is inconsistent with the specified equilibrium conditions. Referring to the Eq. (4.40) expression for $f(E)$, we find that the constancy of the Fermi function in turn requires the Fermi level to be independent of position.

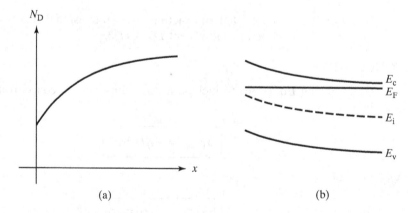

(a)   (b)

Figure 6.14   Nonuniformly doped semiconductor under equilibrium conditions: (a) assumed doping variation with position; (b) corresponding equilibrium energy band diagram.

Turning to the second fact, in Chapter 4 we found that the Fermi level in uniformly-doped n-type semiconductors moved closer and closer to E_c when the donor doping was systematically increased (see Fig. 4.19). Consistent with this observation and as diagramed in Fig. 6.14(b), the distance between E_c and E_F is smaller in regions of higher doping. Band bending is therefore a natural consequence of spatial variations in doping, and

> *a nonzero electric field is established inside nonuniformly doped semiconductors under equilibrium conditions.*

With the preliminary considerations completed, we can now proceed to the derivation proper. Under equilibrium conditions the total carrier currents inside a semiconductor must be identically zero. Still considering a nonuniformly doped semiconductor under equilibrium conditions, and simplifying the presentation by working in only one dimension, we can therefore state

$$J_{N|\text{drift}} + J_{N|\text{diff}} = q\mu_n n\mathscr{E} + qD_N\frac{dn}{dx} = 0 \tag{6.38}$$

However,

$$\mathscr{E} = \frac{1}{q}\frac{dE_c}{dx} \tag{6.39}$$
$$\text{(Same as 4.46)}$$

and

$$n = N_C\mathscr{F}_{1/2}(\eta_c) \tag{6.40}$$
$$\text{(Same as 4.49a)}$$

where $\eta_c = (E_F - E_c)/kT$. Consequently, with $dE_F/dx = 0$ under equilibrium conditions,

$$\frac{dn}{dx} = -\frac{1}{kT}\frac{dn}{d\eta_c}\frac{dE_c}{dx} = -\frac{q}{kT}\frac{dn}{d\eta_c}\mathscr{E} \tag{6.41}$$

Substituting dn/dx from Eq. (6.41) into Eq. (6.38), and rearranging the result slightly, one obtains

$$q\mathscr{E}\left(\mu_n n - \frac{q}{kT}\frac{dn}{d\eta_c}D_N\right) = 0 \tag{6.42}$$

Since $\mathscr{E} \neq 0$ (a consequence of the nonuniform doping), it follows from Eq. (6.42) that

$$\left[\frac{D_N}{\mu_n} = \frac{kT}{q} \frac{n}{dn/d\eta_c} \right] \tag{6.43}$$

This is the generalized form of the Einstein relationship. In the nondegenerate limit, $n \to N_C \exp(\eta_c)$, $n/(dn/d\eta_c) \to 1$, and Eq. (6.43) simplifies to

$$\boxed{\frac{D_N}{\mu_n} = \frac{kT}{q}} \qquad \text{Einstein relationship} \atop \text{for electrons} \tag{6.44a}$$

A similar argument for holes yields

$$\boxed{\frac{D_P}{\mu_p} = \frac{kT}{q}} \qquad \text{Einstein relationship} \atop \text{for holes} \tag{6.44b}$$

Although established assuming equilibrium conditions, we can present more elaborate arguments that show the Einstein relationship to be valid even under non-equilibrium conditions. Eq. (6.43) or the analogous hole relationship must be employed, of course, if the semiconductor is degenerate. Convenient analytical approximations for $n/(dn/d\eta_c) = \mathscr{F}_{1/2}(\eta_c)/\mathscr{F}_{-1/2}(\eta_c)$ are available in the semiconductor literature.[20] Finally, since $kT/q \simeq 0.026$ V at 300 K, in a nondegenerate semiconductor where $\mu_n = 1000$ cm^2/V-sec, $D_N \simeq 26$ cm^2/sec. This provides an idea as to the typical size of the diffusion coefficients.

6.3 EQUATIONS OF STATE

In the preceding two sections and the previous chapter we separately modeled the primary types of carrier action taking place inside a semiconductor. The results from the drift, diffusion, and recombination–generation analyses are combined in this section into an overall mathematical description of the dynamic state within a semiconductor. The "equations of state" thereby established are the basic set of equations that must be solved to determine system variables under arbitary conditions.

6.3.1 Current Equations

Carrier Currents

The electron and hole currents per unit area inside a semiconductor, $\mathbf{J}_N$ and $\mathbf{J}_P$, are obtained by adding the current densities due to drift and diffusion associated with a given carrier. Specifically,

$$\mathbf{J}_P = q\mu_p p \mathscr{E} - qD_P \nabla p \tag{6.45a}$$
$$\updownarrow \text{drift} \qquad \updownarrow \text{diffusion}$$
$$\mathbf{J}_N = q\mu_n n \mathscr{E} + qD_N \nabla n \tag{6.45b}$$

The total current flowing inside a semiconductor under steady-state conditions (or the total *particle* current under arbitrary conditions) is in turn obtained by summing the electron and hole currents:

$$\boxed{\mathbf{J} = \mathbf{J}_N + \mathbf{J}_P} \qquad \text{(steady-state conditions)} \qquad (6.46)$$

Dielectric Displacement Current

It is a well-known fact that current will flow into and out of a capacitor under a.c. and transient conditions. However, there are no particle currents passing through an ideal capacitor. Rather, a change in dielectric polarization within the insulating material maintains current continuity across the capacitor. This change in polarization may be viewed as giving rise to a nonparticle current, the *dielectric displacement current*. In mathematical terms,

$$\mathbf{j}_D = \frac{\partial \mathbf{D}}{\partial t} \qquad (6.47)$$

where $\mathbf{j}_D$ is the displacement current density and $\mathbf{D}$ the dielectric displacement vector.

Under a.c. and transient conditions a change in dielectric polarization also occurs inside semiconductors. The associated dielectric displacement current can, in fact, be the dominant current component in depletion regions. More generally, the displacement current adds to the carrier currents, yielding

$$\boxed{\mathbf{j} = \mathbf{J}_N + \mathbf{J}_P + \frac{\partial \mathbf{D}}{\partial t}} \qquad \text{(a.c. and transient conditions)} \qquad (6.48)$$

A small $\mathbf{j}$ is used here to distinguish the total current density under a.c. and transient conditions from the total current density under steady-state conditions. The dielectric displacement current vanishes, of course, under steady-state conditions. Note that if the semiconductor is a linear dielectric ($\mathbf{D} = K_S \varepsilon_0 \mathbf{\mathscr{E}}$, K_S being a scalar constant), then $\partial \mathbf{D}/\partial t = K_S \varepsilon_0\, \partial \mathbf{\mathscr{E}}/\partial t$.

Quasi-Fermi Levels

Quasi-Fermi levels are conceptual constructs, defined energy levels that can be used in conjunction with the energy band diagram to specify the carrier concentrations inside a semiconductor under nonequilibrium conditions. Relative to the current-equation discussion, the quasi-Fermi level construct also allows one to recast current and other carrier-action relationships into a more compact form.

In the quasi-Fermi level formalism one introduces two energies, F_N, the quasi-Fermi level for electrons, and F_P, the quasi-Fermi level for holes. These energies are *by definition* related to the nonequilibrium carrier concentrations in the same way E_F is related to the equilibrium carrier concentrations. To be specific, under nonequilibrium conditions and assuming the semiconductor to be nondegenerate,

$$n \equiv n_i e^{(F_N - E_i)/kT} \qquad \text{or} \qquad F_N \equiv E_i + kT \ln(n/n_i) \qquad (6.49a)$$

and

$$p \equiv n_i e^{(E_i - F_p)/kT} \qquad \text{or} \qquad F_P \equiv E_i - kT \ln(p/n_i) \qquad (6.49b)$$

Please note that F_N and F_P are totally determined from a prior knowledge of n and p. Moreover, the quasi-Fermi level definitions have been carefully chosen so that when a perturbed system relaxes back toward equilibrium, $F_N \rightarrow E_F$, $F_P \rightarrow E_F$, and Eqs. (6.49) $\rightarrow$ Eqs. (4.57).

To illustrate use of the formalism, let us consider a Si sample doped with $N_D = 10^{14}/\text{cm}^3$ donors and maintained at 300 K under equilibrium conditions. Employing relationships established in Chapter 4, we conclude that $n_0 = 10^{14}/\text{cm}^3$, $p_0 = 10^6/\text{cm}^3$, and $E_F - E_i = kT \ln(N_D/n_i) \simeq 0.24$ eV. The sample is therefore characterized by the equilibrium energy band diagram shown in Fig. 6.15(a). Now suppose the sample is subjected to a uniform perturbation such that $\Delta n = \Delta p = 10^{11}/\text{cm}^3$. Using Eqs. (6.49) with $n = n_0 + \Delta n \simeq n_0$ and $p = p_0 + \Delta p \simeq 10^{11}/\text{cm}^3$, one calculates $F_N - E_i \simeq E_F - E_i$ and $F_P - E_i \simeq -0.060$ eV. Thus the nonequilibrium situation is described by the band diagram shown in Fig. 6.15(b). Note that the mere inclusion of quasi-Fermi levels on any energy band diagram indicates that nonequilibrium conditions prevail. Inspection of Fig. 6.15(b) also indicates at a glance that $p > n_i$; comparison with Fig. 6.15(a) further suggests $n \simeq n_0$.

Having introduced and exhibited the general utility of quasi-Fermi levels, we can finally turn to the development of alternative expressions for $\mathbf{J}_P$ and $\mathbf{J}_N$. Differentiating both sides of Eq. (6.49b) with respect to position, one obtains

$$\nabla p = (n_i/kT)e^{(E_i - F_p)/kT}(\nabla E_i - \nabla F_P) \qquad (6.50a)$$

$$= (qp/kT)\mathscr{E} - (p/kT)\nabla F_P \qquad (6.50b)$$

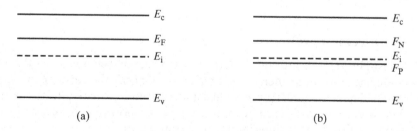

(a)   (b)

Figure 6.15   Sample use of quasi-Fermi levels. Energy band description of the situation inside the text-described Si sample under (a) equilibrium conditions and (b) nonequilibrium conditions.

The identity $\mathscr{E} = \nabla E_c/q = \nabla E_i/q$ [the three-dimensional version of Eq. (6.39)] is employed in progressing from Eq. (6.50a) to Eq. (6.50b). Next, eliminating ∇p in Eq. (6.45a) using Eq. (6.50b) gives

$$\mathbf{J}_P = q(\mu_p - qD_P/kT)p\mathscr{E} + (qD_P/kT)p\nabla F_P \qquad (6.51)$$

From the Einstein relationship, however, $qD_P/kT = \mu_p$. We therefore conclude

$$\boxed{\mathbf{J}_P = \mu_p p\nabla F_P} \qquad (6.52a)$$

Similarly,

$$\boxed{\mathbf{J}_N = \mu_n n\nabla F_N} \qquad (6.52b)$$

Starting with the degenerate analogs of Eq. (6.49), one can show that the same results are obtained even if the semiconductor is degenerate.

Equations (6.52) are sometimes preferred over Eqs. (6.45) in advanced device analyses. Since $\mathbf{J}_P \propto \nabla F_P$ and $\mathbf{J}_N \propto \nabla F_N$, the alternative relationships also lead to a very interesting general interpretation of energy band diagrams containing quasi-Fermi levels. Namely, if the quasi-Fermi levels vary with position ($dF_P/dx \neq 0$ and/or $dF_N/dx \neq 0$), one is informed at a glance that current is flowing inside the semiconductor.

6.3.2 Continuity Equations

Each and every type of carrier action, whether it be drift, diffusion, R–G center recombination, R–G center generation, or some other type of carrier action, gives rise to a change in the carrier concentrations with time. The combined effect of all types of carrier action can therefore be taken into account by equating the overall change in the carrier concentrations per unit time ($\partial n/\partial t$ or $\partial p/\partial t$) to the sum of the $\partial n/\partial t$'s or $\partial p/\partial t$'s due to the individual processes; that is,

$$\frac{\partial n}{\partial t} = \frac{\partial n}{\partial t}\bigg|_{\text{drift}} + \frac{\partial n}{\partial t}\bigg|_{\text{diff}} + \frac{\partial n}{\partial t}\bigg|_{\text{R-G}} + \frac{\partial n}{\partial t}\bigg|_{\substack{\text{other processes}\\ \text{(photogen., etc.)}}} \qquad (6.53a)$$

$$\frac{\partial p}{\partial t} = \frac{\partial p}{\partial t}\bigg|_{\text{drift}} + \frac{\partial p}{\partial t}\bigg|_{\text{diff}} + \frac{\partial p}{\partial t}\bigg|_{\text{R-G}} + \frac{\partial p}{\partial t}\bigg|_{\substack{\text{other processes}\\ \text{(photogen., etc.)}}} \qquad (6.53b)$$

Mathematically, Eqs. (6.53) connect the carrier concentrations at a given point and time with the carrier concentrations at adjacent points and at $\pm\Delta t$. For this reason they are known as the *continuity equations*.

The continuity equations can be written in a more compact form by introducing

$$g_N \equiv \left.\frac{\partial n}{\partial t}\right|_{\substack{\text{other}\\\text{processes}}} \quad ; \quad g_P \equiv \left.\frac{\partial p}{\partial t}\right|_{\substack{\text{other}\\\text{processes}}} \tag{6.54a,b}$$

reintroducing (see Subsection 5.2.2)

$$r_N \equiv -\left.\frac{\partial n}{\partial t}\right|_{R-G} \quad ; \quad r_P \equiv -\left.\frac{\partial p}{\partial t}\right|_{R-G} \tag{6.55a,b}$$

and noting

$$\left.\frac{\partial n}{\partial t}\right|_{\text{drift}} + \left.\frac{\partial n}{\partial t}\right|_{\text{diff}} = \frac{1}{q}\nabla \cdot \mathbf{J}_N \tag{6.56a}$$

$$\left.\frac{\partial p}{\partial t}\right|_{\text{drift}} + \left.\frac{\partial p}{\partial t}\right|_{\text{diff}} = -\frac{1}{q}\nabla \cdot \mathbf{J}_P \tag{6.56b}$$

Eqs. (6.56), which can be established by a straightforward mathematical manipulation, merely state that there will be a change in the carrier concentrations within a given small region of the semiconductor if an imbalance exists between the total carrier currents into and out of the region. Utilizing Eqs. (6.54) through (6.56), we obtain

$$\frac{\partial n}{\partial t} = \frac{1}{q}\nabla \cdot \mathbf{J}_N - r_N + g_N \tag{6.57a}$$

$$\frac{\partial p}{\partial t} = -\frac{1}{q}\nabla \cdot \mathbf{J}_P - r_P + g_P \tag{6.57b}$$

Continuity equations (6.57) are completely general and directly or indirectly constitute the starting point in most device analyses. In computer simulations the continuity equations are often employed directly. The appropriate relationships for r_N, r_P, g_N, and g_P are substituted into Eqs. (6.57), and numerical solutions are sought for $n(x, y, z, t)$ and $p(x, y, z, t)$. In problems where a closed-form type of solution is desired, the continuity equations are typically utilized only in an indirect fashion. The actual starting point in such analyses is a simplified version of the continuity equations to be established in the next subsection.

6.3.3 Minority Carrier Diffusion Equations

The minority carrier diffusion equations, equations extensively employed in device analyses, are derived from the continuity equations by invoking the following set of simplifying assumptions: (1) The particular system under analysis is *one-dimensional*—i.e., all variables are at most a function of just one coordinate (usually the x-coordinate). (2) The analysis is restricted to *minority carriers*. (3) $\mathscr{E} \simeq 0$ in the semiconductor or in regions of the semiconductor subject to analysis. (4) The equilibrium minority carrier concentrations are not a function of position. In other words, $n_0 \neq n_0(x)$, $p_0 \neq p_0(x)$. (5) *Low level injection* conditions prevail. (6) There are no "other processes," except possibly photogeneration, taking place within the semiconductor.

Working on the continuity equation for electrons, we note that

$$\frac{1}{q} \nabla \cdot \mathbf{J}_N \rightarrow \frac{1}{q} \frac{\partial J_N}{\partial x} \qquad (6.58)$$

if the system is one-dimensional. Moreover,

$$J_N = q\mu_n n \mathscr{E} + q D_N \frac{\partial n}{\partial x} \simeq q D_N \frac{\partial n}{\partial x} \qquad (6.59)$$

when $\mathscr{E} \simeq 0$ and one is concerned only with minority carriers. By way of explanation, the drift component can be neglected in the current density expression because $\mathscr{E}$ is small by assumption and minority carrier concentrations are also small, making the $n\mathscr{E}$ product extremely small. (Note that the same argument cannot be applied to majority carriers.) Since, by assumption, $n_0 \neq n_0(x)$, and by definition $n = n_0 + \Delta n$, we can also write

$$\frac{\partial n}{\partial x} = \frac{\partial n_0}{\partial x} + \frac{\partial \Delta n}{\partial x} = \frac{\partial \Delta n}{\partial x} \qquad (6.60)$$

Combining Eqs. (6.58) through (6.60) yields

$$\frac{1}{q} \nabla \cdot \mathbf{J}_N \rightarrow D_N \frac{\partial^2 \Delta n}{\partial x^2} \qquad (6.61)$$

Turning to the remaining terms in the electron continuity equation, the low level injection restriction (5) combined with the minority carrier limitation (2) permits us to employ the specialized relationship for r_N derived in Subsection 5.2.5:

$$r_N = \frac{\Delta n}{\tau_n} \qquad (6.62)$$

In addition, simplification (6) allows us to write

$$g_N = G_L \qquad (6.63)$$

where G_L is the number of electron-hole pairs generated per sec-cm^3 by the absorption of externally introduced photons. It is understood that $G_L = 0$ if the semiconductor is not subject to illumination. Finally, the equilibrium electron concentration is never a function of time, $n_0 \neq n_0(t)$, and we can therefore write

$$\frac{\partial n}{\partial t} = \frac{\partial n_0}{\partial t} + \frac{\partial \Delta n}{\partial t} = \frac{\partial \Delta n}{\partial t} \tag{6.64}$$

Substituting Eqs. (6.61) through (6.64) into the (6.57a) continuity equation, and simultaneously recording the analogous result for holes, one obtains

$$\frac{\partial \Delta n_p}{\partial t} = D_N \frac{\partial^2 \Delta n_p}{\partial x^2} - \frac{\Delta n_p}{\tau_n} + G_L \tag{6.65a}$$

Minority carrier
diffusion equations

$$\frac{\partial \Delta p_n}{\partial t} = D_P \frac{\partial^2 \Delta p_n}{\partial x^2} - \frac{\Delta p_n}{\tau_p} + G_L \tag{6.65b}$$

We have added subscripts to the carrier concentrations in Eqs. (6.65) to remind the user that the equations are valid only for minority carriers, applying to electrons in *p*-type materials and to holes in *n*-type materials.

6.3.4   Equations Summary

For the reader's convenience we have collected in this final subsection the equations routinely encountered in carrier transport and related device analyses. The equations are a repetition of relationships previously established in this chapter, except for Eqs. (6.68) and (6.70), which were introduced in Subsection 4.4.3. Please note that the ρ appearing in Eqs. (6.68) and (6.70) is the charge density (the charge/cm^3), and not the resistivity.

$$\frac{\partial n}{\partial t} = \frac{1}{q} \nabla \cdot \mathbf{J}_N - r_N + g_N \tag{6.66a}$$

$\left(\begin{array}{c} \text{Continuity} \\ \text{equations} \end{array} \right)$

$$\frac{\partial p}{\partial t} = -\frac{1}{q} \nabla \cdot \mathbf{J}_P - r_P + g_P \tag{6.66b}$$

$$\frac{\partial \Delta n_p}{\partial t} = D_N \frac{\partial^2 \Delta n_p}{\partial x^2} - \frac{\Delta n_p}{\tau_n} + G_L \tag{6.67a}$$

$\left(\begin{array}{c} \text{Minority carrier} \\ \text{diffusion equations} \end{array} \right)$

$$\frac{\partial \Delta p_n}{\partial t} = D_P \frac{\partial^2 \Delta p_n}{\partial x^2} - \frac{\Delta p_n}{\tau_p} + G_L \tag{6.67b}$$

$$\nabla \cdot \boldsymbol{\mathscr{E}} = \rho/K_S \varepsilon_0 \quad \text{(Poisson's equation)} \tag{6.68}$$

where

$$\mathbf{J}_N = q\mu_n n\boldsymbol{\mathscr{E}} + qD_N \nabla n = \mu_n n \nabla F_N \tag{6.69a}$$

$$\mathbf{J}_P = q\mu_p p\boldsymbol{\mathscr{E}} - qD_P \nabla p = \mu_p p \nabla F_P \tag{6.69b}$$

$$\rho = q(p - n + N_D^+ - N_A^-) \tag{6.70}$$

$$\mathbf{J} = \mathbf{J}_N + \mathbf{J}_P \tag{6.71}$$

$$\mathbf{j} = \mathbf{J}_N + \mathbf{J}_P + \frac{\partial \mathbf{D}}{\partial t} \tag{6.72}$$

and

$$r_N \equiv -\left.\frac{\partial n}{\partial t}\right|_{R-G} \quad ; \quad r_P \equiv -\left.\frac{\partial p}{\partial t}\right|_{R-G} \tag{6.73a,b}$$

$$g_N \equiv \left.\frac{\partial n}{\partial t}\right|_{\substack{\text{other} \\ \text{processes}}} \quad ; \quad g_P \equiv \left.\frac{\partial p}{\partial t}\right|_{\substack{\text{other} \\ \text{processes}}} \tag{6.74a,b}$$

REFERENCES

[1] C. Jacoboni, C. Canali, G. Ottaviani, and A. A. Quaranta, "A Review of Some Charge Transport Properties of Silicon," Solid-State Electronics, 20, 77 (1977).

[2] P. M. Smith, J. Frey, and P. Chatterjee, "High-Field Transport of Holes in Silicon," Appl. Phys. Letts., 39, 332 (Aug., 1981).

[3] J. S. Blakemore, "Semiconducting and Other Major Properties of Gallium Arsenide," J. Appl. Phys., 53, R123 (Oct., 1982).

[4] W. Walukiewicz, L. Lagowski, L. Jastrzebski, M. Lichtensteiger, and H. C. Gatos, "Electron Mobility and Free Carrier Absorption in GaAs: Determination of the Compensation Ratio," J. Appl. Phys., 50, 899 (Feb., 1979).

[5] J. Bardeen and W. Shockley, "Deformation Potentials and Mobilities in Nonpolar Crystals," Phys. Rev., 80, 72 (1950).

[6] E. M. Conwell and V. F. Weisskopf, "Theory of Impurity Scattering in Semiconductors," Phys. Rev., 77, 388 (1950).

[7] P. P. Debye and E. M. Conwell, "Electrical Properties of n-type Ge," Phys. Rev., 93, 693 (Feb., 1954).

[8] S. S. Li and W. R. Thurber, "The Dopant Density and Temperature Dependence of Electron Mobility and Resistivity in n-type Silicon," Solid-State Electronics, 20, 609 (1977).

[9] S. S. Li, "The Dopant Density and Temperature Dependence of Hole Mobility and Resistivity in Boron Doped Silicon," Solid-State Electronics, 21, 1109 (1978).

[10] L. C. Linares and S. S. Li, "An Improved Model for Analyzing Hole Mobility and Resistivity in p-type Silicon Doped with Boron, Gallium, and Indium," J. Electrochem. Soc., *128*, 601 (March, 1981).

[11] D. M. Caughey and R. F. Thomas, "Carrier Mobility in Silicon Empirically Related to Doping and Field," Proc. IEEE, *55*, 2192 (Dec., 1967).

[12] N. D. Arora, J. R. Hauser, and D. J. Roulston, "Electron and Hole Mobilities in Silicon as a Function of Concentration and Temperature," IEEE Trans. on Electron Devices, *ED-29*, 292 (Feb., 1982).

[13] G. Baccarani and P. Ostoja, "Electron Mobility Empirically Related to the Phosphorus Concentration in Silicon," Solid-State Electronics, *18*, 579 (1975).

[14] C. M. Maziar and M. S. Lundstrom, "Caughey-Thomas Parameters for Electron Mobility Calculations in GaAs," Electronics Letters, *22*, 565 (1986).

[15] S. M. Sze, *Physics of Semiconductor Devices*, 2nd edition, John Wiley & Sons, New York, 1981: See Fig. 21, p. 32.

[16] C. Bulucea, "Recalculation of Irvin's Resistivity Curves for Diffused Layers in Silicon Using Updated Bulk Resistivity Data," Solid-State Electronics, *36*, 489 (1993).

[17] W. R. Runyan and T. J. Shaffner, *Semiconductor Measurements and Instrumentation*, 2nd edition, McGraw-Hill Book Co., Inc., New York, 1998.

[18] L. J. van der Pauw, "A Method of Measuring Specific Resistivity and Hall Effect of Discs of Arbitrary Shape," Philips Research Reports, *13*, 1 (1958).

[19] R. D. Present, *Kinetic Theory of Gases*, McGraw-Hill Book Co. Inc., New York, 1958.

[20] N. G. Nilsson, "Empirical Approximations Applied to the Generalized Einstein Relation for Degenerate Semiconductors," Phys. Stat. Sol. (a), *50*, K43 (1978).

PROBLEMS

6.1 Answer the following questions as concisely as possible.

(a) Figure 6.3 was constructed from v_d versus $\mathscr{E}$ data derived from high-purity (lightly doped) samples. How would the low-field region of the plots be modified if the v_d versus $\mathscr{E}$ measurements were performed on samples with higher doping concentrations? Explain.

(b) It is determined that $\mu_{Ln} = 1360$ cm^2/V-sec and $\mu_{In} = 2040$ cm^2/V-sec in a Si sample at 300 K. Assuming phonon and impurity ion scattering to be dominant, compute the expected value of μ_n employing (i) Matthiessen's rule and (ii) Eq. (6.8).

(c) Assuming that $N_A = N_D = 0$, compute the resistivity of (i) intrinsic Si and (ii) intrinsic GaAs at 300 K.

(d) Sketch the approximate variation of the hole diffusion coefficient (D_P) versus doping (N_A) in silicon at room temperature. Also explain how you arrived at your answer.

(e) For temperatures T near room temperature, sketch the expected form of a log-log D_N versus T plot appropriate for *lightly* doped *n*-type silicon. Record the reasoning leading to your sketch.

6.2 (a) Paralleling the text derivation of Eq. (6.27a), derive Eq. (6.27b).

(b) Confirm that Eq. (6.52a) is valid for degenerate semiconductors; i.e., rederive Eq. (6.52a) taking the semiconductor to be degenerate.

(c) Derive Eq. (6.56b).

6.3 Relative to the physical interpretation of the mobility it was noted in a footnote that $\mu = q\tau/m^*$. Later, in deriving the current due to diffusion, we equated (quoting the exact result) $D = \bar{v}l/3$. Moreover, in Subsection 5.2.6 we pointed out that $m^*v_{th}^2/2 = 3kT/2$, where v_{th} is the thermal velocity or average velocity under equilibrium conditions. Show that the Einstein relationship follows directly from the cited relationships.

6.4 The energy band diagram pictured in Fig. P6.4 characterizes a Si sample maintained at room temperature. Note that $E_F - E_i = E_G/4$ at $x = \pm L$ and $E_i - E_F = E_G/4$ at $x = 0$.

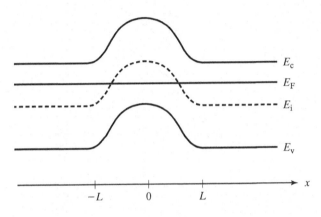

Figure P6.4

(a) The semiconductor is in equilibrium. How does one deduce this fact from the given energy band diagram?

(b) What is the electron current density (J_N) and hole current density (J_P) at $x = \pm L/2$?

(c) Roughly sketch n and p versus x inside the sample.

(d) Is there an electron diffusion current at $x = \pm L/2$? If there is a diffusion current at a given point, indicate the direction of *current* flow.

(e) Sketch the electric field ($\mathscr{E}$) inside the semiconductor as a function of x.

(f) Is there an electron drift current at $x = \pm L/2$? If there is a drift current at a given point, indicate the direction of current flow.

6.5 The energy band diagram characterizing a uniformly doped Si sample maintained at room temperature is pictured Fig. P6.5.

(a) Sketch the electron and hole concentrations (n and p) inside the sample as a function of position.

(b) Sketch the electron and hole diffusion current densities ($J_{N|diff}$ and $J_{P|diff}$) inside the sample as a function of position.

(c) Sketch the electric field ($\mathscr{E}$) inside the semiconductor as a function of position.

(d) Sketch the electron and hole drift current densities ($J_{N|drift}$ and $J_{P|drift}$) inside the sample as a function of position.

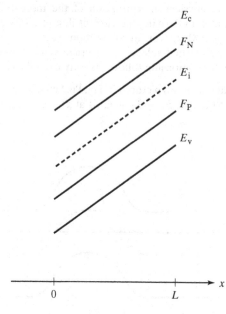

Figure P6.5

6.6 A portion $(0 \leq x \leq L)$ of a Si sample, uniformly doped with $N_D = 10^{15}/cm^3$ donors and maintained at room temperature, is subject to a *steady state* perturbation such that

$$n \cong N_D$$

$$p = n_i(1 - x/L) + n_i^2/N_D \quad ...0 \leq x \leq L$$

Since $n \cong N_D$, it is reasonable to assume $\mathscr{E} \cong 0$ in the $0 \leq x \leq L$ region. Given $\mathscr{E} \cong 0$, sketch the energy band diagram for the perturbed region specifically including E_c, E_i, E_v, F_N, and F_P on your diagram.

6.7 A laser beam striking a uniformly doped p-type bar of silicon maintained at room temperature causes a steady state excess of $\Delta n_p = 10^{11}/cm^3$ electrons at $x = 0$. Note that the laser-induced photogeneration only occurs at $x = 0$. As pictured in Fig. P6.7, the bar extends from $x = -L$ to $x = +L$ and $\Delta n_p(-L) = \Delta n_p(+L) = 0$. $N_A = 10^{16}/cm^3$ and $\mathscr{E} \cong 0$ inside the bar.

(a) What are the dominant physical processes that determine the steady-state excess electron concentration $[\Delta n_p(x)]$ in the regions of the bar removed from $x = 0$? Your choices are drift, diffusion, recombination, and generation.

(b) Sketch the expected general form of $\Delta n_p(x)$ inside the bar $(-L \leq x \leq L)$ under steady state conditions.

(c) Does low level injection exist under steady state conditions? Explain.

(d) Reduced to the simplest possible form, write down the equation that must be solved to determine $\Delta n_p(x)$ for $0 < x \leq L$.

(e) What is the general solution to the part (d) equation?

(f) What are the boundary conditions that must be applied in solving the part (d) equation to determine the solution constants?

(g) Complete the solution by applying the boundary conditions to obtain $\Delta n_p(x)$ for $0 < x \le L$.

(h) What is the limit of the part (g) solution if $L \to \infty$?

(i) What is the limit of the part (g) solution if $L \ll L_N$, where $L_N \equiv \sqrt{D_N \tau_n}$ is known as the minority carrier diffusion length?

Laser Beam

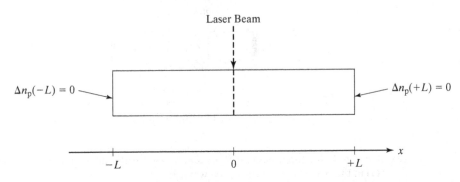

$\Delta n_p(-L) = 0$

$\Delta n_p(+L) = 0$

$-L$ $\qquad$ 0 $\qquad$ $+L$ $\qquad$ x

Figure P6.7

6.8 A short n-type GaAs bar of length L (see Fig. P6.8) is subject to a perturbation such that, under steady-state conditions,

$$\Delta p_n(x) = \Delta p_{n0}(1 - x/L) \quad \ldots 0 \le x \le L$$

The GaAs bar is uniformly doped with $N_D = 10^{16}/\text{cm}^3$ donors *and* $N_A = 5 \times 10^{15}/\text{cm}^3$ acceptors, $\Delta p_{n0} = 10^{10}/\text{cm}^3$, and $T = 300$ K.

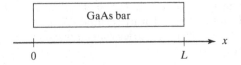

GaAs bar

0 $\qquad$ L $\qquad$ x

Figure P6.8

(a) Characterize the bar under *equilibrium* conditions by providing numerical values for (i) n_i, (ii) n_0, and (iii) p_0.

(b) Does the cited perturbed state correspond to a "low level injection" situation? Explain.

(c) For the given perturbation it is reasonable to assume $\mathscr{E} \simeq 0$ everywhere in the bar. Given $\mathscr{E} \simeq 0$, sketch the energy band diagram for $0 \le x \le L$ specifically including E_c, E_i, E_v, F_N, and F_P on your diagram. Only the rough positionings of F_N and F_P are required.

(d) (i) There must be a hole diffusion current in the bar. Explain why in words.

(ii) The hole drift current should be negligible compared to the hole diffusion current. Explain why.

(iii) Establish an expression for the hole current density.

(e) Show that the $\Delta p_n(x)$ quoted in the statement of the problem can be obtained by assuming R–G center recombination–generation and "other processes" are negligible inside the bar, solving the simplified minority carrier diffusion equation, and applying the boundary conditions $\Delta p_n(0) = \Delta p_{n0}$, $\Delta p_n(L) = 0$.

6.9 The $x = 0$ to $x = L$ section of a Si wafer maintained at room temperature is nondegenerately doped with $N_D \gg n_i$ donors/cm^3. Moreover, the semiconductor is subjected to a steady-state perturbation that makes both n and p much less than n_i everywhere in the $x = 0$ to $x = L$ section of the wafer (see Fig. P6.9).

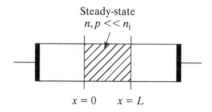

Steady-state
$n, p \ll n_i$

$x = 0 \qquad x = L$

Figure P6.9

(a) If $E_T' = E_i$ for the dominant R–G center, what is the net steady-state generation rate (G) in the perturbed section of the wafer?

(b) Can the perturbation here be classified as "low level injection"? Explain.

(c) Is the perturbed region charge neutral? If not, what is ρ (the charge/cm^3) in the perturbed region?

(d) Referring to Poisson's equation, assuming an $\mathscr{E}$-field only in the x-direction, and taking $\mathscr{E} = 0$ at $x = 0$, determine $\mathscr{E}$ versus x for $0 \leq x \leq L$.

(e) Considering your $\mathscr{E}$-field result, make a *rough* sketch of the energy band diagram (E versus x) within the perturbed region.

(f) Could the minority carrier diffusion equation for holes be utilized in an analysis treating the $x = 0$ to $x = L$ region? Explain.

(g) Assuming $J_P = 0$ at $x = 0$, $J_N = 0$ at $x = L$, and only x-direction current flow, determine the total current density (J) inside the perturbed region.

HINT: Write down the carrier *continuity* equations appropriate for the perturbed region. Also assume $g_N = g_P = 0$.

6.10 Consider a nondegenerate, uniformly doped, p-type semiconductor sample maintained at room temperature. At time $t = 0$ a pulse-like perturbation causes a *small* enhancement of the MAJORITY-carrier hole concentration at various points inside the sample. We wish to show that the perturbation in the hole concentration [$\Delta p(t)$] will decay exponentially with time and that the decay is characterized by a time constant $\tau = \varepsilon/\sigma = K_S \varepsilon_0 / q\mu_p N_A$. τ is referred to as the *dielectric relaxation time*—the time it takes for majority carriers to rearrange in response to a perturbation.

(a) Write down the continuity equation for holes. (Why not write down the minority carrier diffusion equation for holes?)

(b) Write down the properly simplified form of the hole continuity equation under the assumption that R–G center recombination–generation and all "other processes" inside the sample have a negligible effect on $\Delta p(t)$.

(c) Next, assuming that diffusion at all points inside the sample is negligible compared to drift, write down the appropriate expression for $\mathbf{J}_P$. After further simplifying $\mathbf{J}_P$ by noting $p = N_A + \Delta p \simeq N_A$, substitute your $\mathbf{J}_P$ result into the part (b) result.

(d) Write down Poisson's equation and explicitly express ρ (the charge density) in terms of the charged entities inside the semiconductor. Simplify your result, noting that $N_A \gg N_D$ and $p \gg n$ for the given sample and conditions.

(e) To complete the analysis:
   (i) Combine the part (c) and (d) results to obtain a differential equation for p.
   (ii) Let $p = N_A + \Delta p$.
   (iii) Solve for $\Delta p(t)$. As stated earlier, $\Delta p(t)$ should be an exponential function of time characterized by a time constant $\tau = \varepsilon/\sigma$.

(f) Compute τ for $N_A = 10^{15}/cm^3$ doped silicon maintained at room temperature.

Index